Parallel Programs for the Transputer

Ronald S. Cok

Parallel Programs for the Transputer

Parallel Programs for the Transputer

Ronald S. Cok

Eastman Kodak Company

Prentice Hall
Englewood Cliffs, New Jersey 07632

Library of Congress Cataloging-in-Publication Data

Cok, Ronald S.
  Parallel programs for the transputer / Ronald S. Cok.
    p.   cm.
  Includes bibliographical references and index.
  ISBN 0-13-651480-4
  1. Transputers--Programming. 2. Parallel programming (Computer science)   I. Title.
QA76.6.C6245 1991
005.26--dc20   90-37870
                                                                   CIP

Editorial/production supervision: Harriet Tellem
Cover design: Bruce Kenselaar
Manufacturing buyer: Kelly Behr

© 1991 by Prentice-Hall, Inc.
A division of Simon & Schuster
Englewood Cliffs, New Jersey 07632

The publisher offers discounts on this book when ordered in bulk quantities. For more information, write:

  Special Sales/College Marketing
  Prentice-Hall, Inc.
  College Technical and Reference Division
  Englewood Cliffs, New Jersey 07632

occam is a registered trademark of the INMOS Corp.

All rights reserved. No part of this book may be
reproduced, in any form or by any means,
without permission in writing from the publisher.

Printed in the United States of America
10 9 8 7 6 5 4 3 2 1

ISBN 0-13-651480-4

Prentice-Hall International (UK) Limited, *London*
Prentice-Hall of Australia Pty. Limited, *Sydney*
Prentice-Hall Canada Inc., *Toronto*
Prentice-Hall Hispanoamericana, S.A., *Mexico*
Prentice-Hall of India Private Limited, *New Delhi*
Prentice-Hall of Japan, Inc., *Tokyo*
Simon & Schuster Asia Pte. Ltd., *Singapore*
Editora Prentice-Hall do Brasil, Ltda., *Rio de Janeiro*

Dedicated to

Sueanne

for her patience, perfection, and other things beginning with p.

Contents

	Preface	xi
Chapter 0	**Introduction**	**1**
	Parallel Computer Efficiency	2
	Transputers	8
	occam	11
	SEQ	12
	PAR	12
	CHAN	14
	ALT	14
	Replicated Structures	16
	Control Structures	16
	Debugging	19
	Miscellaneous	20
	In Summary	21
Chapter 1	**SISD, SIMD, MIMD, and All That**	**23**
	Multiple Data Paths	25
	Multiple Instruction Paths	26
	Shared Memory	26
	Distributed Memory	28
	Architectural Issues	31
	In Summary	33
Chapter 2	**Architectures**	**35**
	Configuration Descriptions	36
	Networks	39
	Irregular Networks	39
	Rings	42
	Toroids	44
	Hypercubes	47
	Ternary Trees	51
	In Summary	56

Contents

Chapter 3	**Processor Farms**	**57**
	A Simple Processor Farm	59
	Efficiency Concerns	60
	Large Processor Farms	63
	Storage and Communication Issues	69
	A Real-World Example	70
	Efficiency Measurements	83
	In Summary	86
Chapter 4	**Pipeline Processing**	**89**
	Program Issues	89
	Pipeline Efficiency	91
	A Pipeline Example	94
	Communication Methods	96
	Single Buffering	98
	Double Buffering	99
	Triple Buffering	101
	A Buffered Pipeline Test	103
	Multidimensional Pipelines	106
	In Summary	111
Chapter 5	**Data Parallelism**	**113**
	Program Issues	113
	Data Distribution	115
	Loading Data	119
	A Simple Load Routine	119
	A Fast Load Routine	121
	Bidirectional Loading	124
	Performance Measures	127
	Sampling	128
	Expanding Data Sets	130
	One-Dimensional Expansion	131
	Two-Dimensional Expansion	131
	Performance Issues	135
	Communicating Data Sets	137
	Shifting	137
	Performance Issues	141
	An Efficiency Comparison	142
	In Summary	144

Chapter 6	**Deadlock-Free Routing**	**145**
	Program Issues	146
	One-Way Virtual Channels	150
	A One-Way Ring Router	156
	Two-Way Virtual Channels	160
	A Two-Way Ring Router	163
	A Four-Way Toroidal Router	165
	Performance Comparisons	172
	In Summary	173
Chapter 7	**Worms**	**175**
	Searching Strategies	176
	Bootstrapping a Processor	183
	A Simple, Sequential Worm	185
	A Simple, Parallel Worm	193
	A Robust, Exploratory Worm	199
	In Summary	211
Chapter 8	**Real-Time Processing**	**213**
	Interrupt Handlers	214
	A Simple Handler	214
	A Buffered Handler	215
	A FIFO Handler	217
	Performance Comparisons	223
	In Summary	226
	Bibliography	**227**
	Index	**231**

Preface

Parallel computing has come of age and is moving into the mainstream of the computing world. As this technology matures, the variety and number of parallel computers are growing rapidly. Multiple-instruction, multiple-data machines, which use many communicating processors, are one of the several types of parallel computers available today and are becoming increasingly popular. The transputer, a single-chip microcomputer developed by the Inmos Corp., was the first of these communicating processors to be built into a single, integrated circuit, and to be supported within a complete parallel computing environment. This environment includes not only internal hardware support for parallelism within the transputer itself, but also a parallel programming language, occam, and a supporting development system.

One of the truly enjoyable aspects of parallel computing with transputers is the variety of ways, and the ease with which, transputers can be used to build different parallel computers. In this book, I explore and illustrate this diversity. In addition to a discussion of parallel computing approaches and examples of a variety of computer architectures, six programming methodologies for parallel computing are presented. Each of these methodologies is developed with numerous programs illustrating techniques ranging from the simple to the complex. The techniques are compared, the various practical trade-offs between them examined, and their relative performance and efficiency measured. Every complete program presented is taken directly from a working example.

My goal is to illustrate and educate, not to create performance benchmarks. None of the example programs are optimized for execution speed, and any performance measures are presented solely for comparison with other examples. Indeed, the programs are all compiled with the error checking compiler switches turned on. This may increase one's confidence in a program but certainly does not enhance the program's performance.

It is most definitely *not* my expectation that these programming examples are "perfect," or even unique. Programs can often be written in a half dozen different ways, and the examples in this book are no exception. The programs presented here are used to illustrate, explain, analyze, and compare methods that many people can readily program themselves.

This book will be useful to both the novice and the experienced parallel programmer for the transputer, although a rudimentary knowledge of occam and transputers will be very helpful. Novice users will benefit from the introduction of pro-

Preface

gramming approaches and their illustration. Experienced programmers will find the more complex examples and their comparisons with simpler routines useful and enlightening. In addition, readers familiar with some of the topics may not have explored all of the different methodologies presented. After a careful study of this book, any reader will be thoroughly familiar with a broad variety of parallel programming techniques for transputers, and will understand the system trade-offs involved in using them.

Although some readers may study chapters selectively, the topics do follow a logical developmental order, with the more difficult programming techniques presented later in the book. This developmental order is found within each chapter as well as in the progression of chapters. Experienced programmers may wish to proceed directly to the more advanced material.

The first two chapters, 0 and 1, are primarily introductory in nature. In these chapters is presented an overview of parallel computing efficiency, transputer systems, occam, and of general classes of parallel computers. The first part of Chapter 2 is also introductory and contains a discussion of configuration descriptions; in the second part, a variety of architectures is described and configured. The remaining chapters continue with the actual presentation of various programming methods, together with an analysis of the efficiency of the techniques illustrated.

A good book is never written in isolation, and this book is no exception. I wish to express my appreciation to those colleagues whose suggestions or ideas contributed to the many examples in this book. Thanks are also due to my patient grammarian and wife, Sueanne, and to the editors and reviewers, especially David Cok, whose corrections, additions, and deletions did so much to improve and clarify the writing. Any remaining faults are my own.

Ronald S. Cok

Parallel Programs for the Transputer

Chapter 0

Introduction

Parallel processing is one of the most promising and fascinating fields in computer science today. Computers using parallel processors can provide high performance at a reasonable cost. As a result of new advances in integrated circuit technology and software development, it is now possible to build useful, powerful, and inexpensive parallel computers. One of the most intriguing advances has been the development of the transputer, a single-chip computer developed by Inmos Ltd. The word *transputer* is a combination of the words *transistor* and *computer* and is meant to imply that parallel computers can be built of transputers just as traditional computers are built of transistors.

The advent of the transputer and its gradual acceptance by the computing world have made it possible to experiment widely with parallel architectures and programming methodologies with a minimum of effort. This book introduces the different architectures found in parallel computers and presents in detail programming methodologies useful for transputer-based computers of various architectures. One of the strengths of the transputer approach to computer design and programming is the variety of techniques and implementations that can be used for different applications. Many parallel computers using different architectures can be designed around the transputer, and each is likely to require a different parallel programming approach.

The programming approaches presented here are not always unique to parallel processing and may have been developed in other contexts as well, but the techniques described are very useful for transputer-based parallel computers. Some of the methods are simple and straightforward and have undoubtedly been developed by many other parallel programmers. Other methods are less common. In either case, it is the author's hope that a clear exposition of a variety of parallel programming methods will be useful to both the novice and experienced parallel programmer. Most of the techniques discussed have been implemented by the author in real applications. Others are drawn from interesting and useful examples of common computing tasks.

Transputer-based parallel computers exhibit a diversity of network architectures, both switched and unswitched. Obviously, how any program will be written for a transputer network will depend heavily on the interconnection scheme used in the system. The techniques presented are useful for a variety of networks, but the emphasis is on unswitched transputer networks which are simply connected with links. The programming examples for each parallel processing methodology are developed over several levels of complexity and are described together with comments about the strengths, drawbacks, and difficulties of each parallel programming technique.

In the remainder of this chapter are presented a discussion of efficiency in parallel computers, a brief description of the transputer and the systems used by the author to develop the examples, and a short tutorial on the occam† programming language. This discussion is a rudimentary introduction to the transputer and occam; a basic knowledge of occam, the parallel programming language used for all of the examples, is necessary for understanding the programs.

Parallel Computer Efficiency

Efficiency measures are important for the effective use of any computer. In a traditional computer, efficiency is largely a matter of carefully using the cache and memory storage and using good programming methods. In addition to dealing with these traditional concerns, a parallel programmer must effectively use every processor in a network so that no processor is idle while other processors work and so that no work is done more than once. Achieving this can be very difficult.

A simple metric which will be used throughout this book for calculating the efficiency of a program on a parallel computer compares the time required to run a program on a single processor to the time required to run the program on a parallel computer:

$$\text{Efficiency} = \frac{\text{Time on sequential computer}}{\text{Time on parallel computer} \times \text{Number of processors}}$$

If a program takes ten seconds to run on one processor and one second on ten processors, the efficiency is 100%. If the program takes two seconds on ten processors the efficiency is 50%.

The relative efficiency between two parallel processors can be found by dividing the efficiency of one by the efficiency of the other. This relative efficiency can be reduced to:

$$\text{Efficiency}\frac{A}{B} = \frac{\text{Time on parallel computer } B \times \text{Number of processors in } B}{\text{Time on parallel computer } A \times \text{Number of processors in } A}$$

---

† occam is a trademark of the INMOS Group of Companies, and is used in lower case to distinguish it from the 13th century philosopher, Sir William of Occam, for whom the programming language is named.

A second common performance measure for parallel computers is the speedup. The speedup of a parallel computer is the ratio of its processing time to that of a sequential, single-processor computer for a given problem:

$$\text{Speedup} = \frac{\text{Time on sequential computer}}{\text{Time on parallel computer}}$$

The speedup of a parallel computer compared to a single processor is also the efficiency of the parallel computer times the number of processors in it.

All other things being equal, the efficiency of a parallel computer can never truly be greater than 100%. Efficiency greater than 100% is called superlinear. It is possible that a multiprocessor system will appear to exhibit superlinear efficiency, but closer examination of the system will show that this apparent efficiency is due either to subtle program differences between the multiprocessor and single processor systems or differences within the systems themselves. We can easily see the impossibility of superlinear efficiency by observing that a single processor can perform sequentially the same operations that a parallel processor does in parallel. The sum of the time each processor takes cannot be greater than the time a single, sequential processor needs. It is not difficult, however, to create situations in which *not* all things are equal, situations in which a parallel system seems to exhibit superlinear efficiency.

As an example of apparent superlinear efficiency in a parallel computer, consider a single processor with a given amount of memory and a two-processor parallel machine with that same amount of memory in each processor. Both processors have access to a slow, very large memory (such as a magnetic disk) in which data is stored (Fig. 0-1). If the data to be processed can be divided into two parts, each of which must be repeatedly accessed, and if the data set is twice as large as each processor's individual memory, the single processor will have to fetch data repeatedly from the slower storage. In contrast, the parallel machine can read the data once, storing one part of the data set in one processor and the rest in the other. Since the two-processor computer only reads the data from the slow disk once, it can process the data at more than twice the speed of the single processor computer, giving the impression of superlinear efficiency. In this case, however, not all things are equal in these two systems because the parallel *computer* has twice the memory of the single processor, even though each *processor* in both computers has the same amount of memory.

A second, simple example of apparent superlinear speedup readily demonstrable on transputer systems involves the use of program loops. If a given task requires that an operation be repeated ten times, a one-processor system might perform the task by executing a loop ten times. In contrast, a ten-processor system will perform the task with a single step in each processor. Since the single processor will encounter program overhead in executing the loop which the parallel processor

Introduction   *Chapter 0*

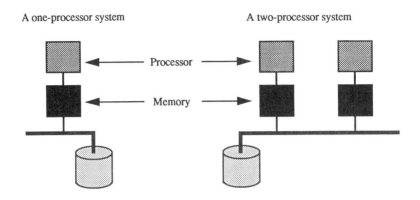

Figure 0-1
Computer systems with memory and disk storage

will not, the ten-processor system will perform the task more than 10 times as fast as the single processor. Of course, if the program loop in the single processor is unrolled, the discrepancy disappears.

In actual practice, apparent superlinear speedup is rarely experienced since parallel programs encounter other kinds of overhead not found in sequential programs. For example, in order to coordinate their work, the processors in a parallel computer must communicate. This communication takes time, time which a sequential, single-processor computer does not spend. The result is that parallel computers are never 100% efficient. Typically, the more communication the processors must do, the worse the efficiency becomes. To help alleviate this problem, transputers have been built to perform communication at the same time as processing. This capability reduces the overhead but does not eliminate it, since any communication must at least be set up by the processor.

Figure 0-2 is a typical logarithmic graph of the efficiency of a parallel computer as the number of processors grows. If there are more processors in a network, more communication is needed while the amount of work to be accomplished remains the same. Thus the relative efficiency of the processor network becomes worse as the number of processors increases. At some point the work cannot be further subdivided among more processors and, as more processors are added, the additional communication overhead actually may cause the overall performance to decrease. At this point, the processing can no longer speed up as the communication overhead continues to rise. At best, the performance will asymptotically approach a limit.

To make this point more concretely, let us consider the hypothetical problem of convolving a 128-element vector with the small two-element kernel [1,1]. A sin-

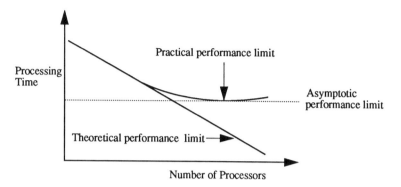

Figure 0-2
Performance limit for a parallel processor (logarithmic scales)

```
[128]INT a:
INT temp:
SEQ
  temp:=a[0]
  SEQ i=0 FOR 126
    SEQ
      a[i]:=a[i]+a[i+1]
  a[127]:=a[127]+temp
```

Figure 0-3
Convolution listing for hypothetical performance demonstration

gle processor must perform 128 adds to calculate the convolution. Figure 0-3 lists a simple program which does this. If we assume that an add can be done in one cycle, 128 cycles are required to convolve the vector. In contrast, a parallel computer with N processors arranged in a ring will have $128 / N$ elements in each processor and require $128 / N$ cycles. However, the last element in each processor must be combined with the first element in the processor to the right in order to perform the two-element convolution. This neighboring element can be obtained if each processor performs an input from the right and an output to the left. A simple program to do this is listed in Fig. 0-4. Let us assume that this communication also requires one cycle to complete. If $N = 2$, the total work will then require 65 cycles; if $N = 4$, 33 cycles are needed; and so on. For N greater than 128, some processors will have no work, and the performance of the system will no longer improve. Figure 0-5 is a graph of the performance of our hypothetical convolution; the efficiency of the calculation is shown in Fig. 0-6. (The simple convolution program listed in Fig. 0-4 works only when num.procs is a power of two, and will not work with more than 128 processors.) From an examination of these figures, we can see that efficiency problems begin occurring when the amount of communication becomes significant compared to the amount of processing. While it is true that in transputer systems, and in this example, the communication can be done concurrently with the processing, thereby reducing the communication overhead, the communication overhead often *does* prove to be a very significant factor in the total system efficiency.

Although this simple vector convolution is an artificial example, it serves to illustrate a basic cause of inefficiency in parallel machines. Real examples using real programs are presented in the following chapters.

```
[128]INT a:
size IS 128/num.procs:
PAR j=0 FOR num.procs
  local IS [a FROM j*size FOR size]:
  SEQ
    PAR
      left! local[0]
      right? data
    SEQ i=0 FOR (size-1)
      local[i]:=local[i]+local[i+1]
    local[size-1]:=local[size-1]+data
```

Figure 0-4
Convolution program for a parallel processor

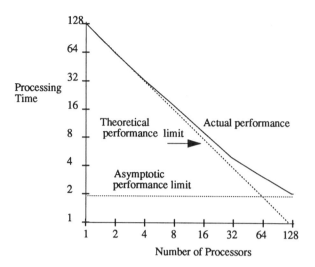

Figure 0-5
Convolution performance for different network sizes

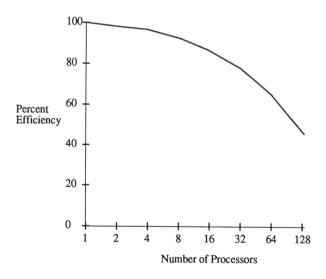

Figure 0-6
Convolution efficiency for different network sizes

Introduction

Transputers

The transputer is a single-chip microprocessor developed by Inmos Ltd. It is the first microcomputer designed specifically for use in parallel processing systems. Among its distinguishing features are special hardware for context switching between parallel processes on a single transputer processor; point-to-point communication links for connecting two processors together; special direct memory access hardware to move data quickly into and out of the links; and an on-chip memory array. All of these features contribute to the efficient implementation of parallel processing tasks.

Figure 0-7 is a diagram of a generic transputer microprocessor. Several different transputer microprocessors actually exist, all of which share the special features listed above. The actual size of the memory array, the number of links, and the structure of the processing units and external memory interface differ among the various transputers available. This particular representation was drawn from the Inmos transputer documentation.† Future transputer products may have hardware features different from those described in this book. In particular, the programs discussed in this book, do not assume the use of virtual or monitor link hardware. The programs are based on the T8xx and earlier series of transputer products.

The transputer links are bidirectional and can pass data both into and out of the processors to which they are connected. Although the link bandwidth in either direction is the same, when the links are used to simultaneously pass data in both directions the overall throughput is not twice the throughput possible when data is passed in only one direction. The transputers used for the demonstrations discussed in this book are capable of passing 1.8 MBytes per second in one direction at a time, and 2.4 MBytes in both directions simultaneously. Therefore, a bidirectional communication can pass only 33% more data than a communication which sends data in one direction.

All of the examples in this book were tested using networks of transputers connected by their links. The architectural structure of the networks is shown in Fig. 0-8. Networks of different sizes are actually used for the various examples, but all of the examples use the same toroidal architectural configuration with an additional processor inserted in one row. The additional processor also connects to the host computer responsible for file handling, display, and keyboard input. The networks themselves consist of T800 transputers running at 20 MHz. In order to increase the overall memory available, an additional 1 MByte of relatively slow, five-cycle dynamic RAM is connected to each processor.

The architecture shown in Fig. 0-8 can be used to implement ring networks, as well as grids and toroids. Subsets of this network are used in different ways to create the various examples and demonstrations presented.

---

†. INMOS Limited, *Transputer Reference Manual*, (Hemel Hempstead, U.K.: Prentice Hall International (U.K.) Ltd., 1988), p. 46.

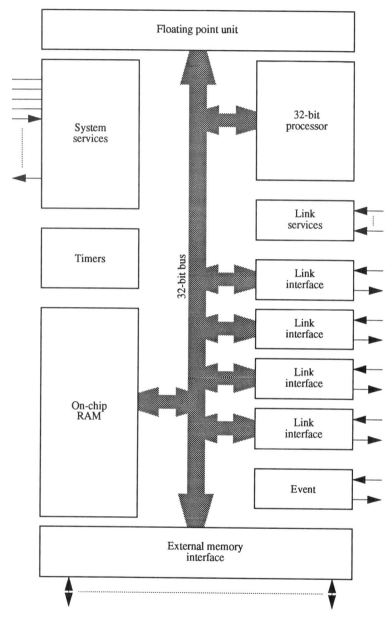

Figure 0-7
Block diagram of a transputer

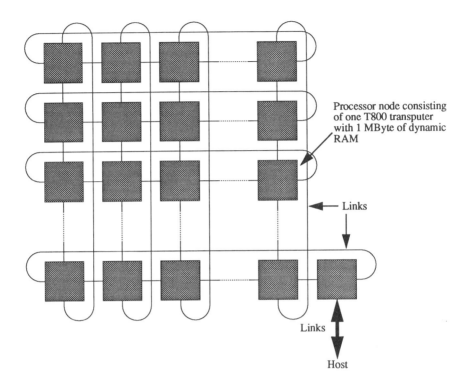

Figure 0-8
Architecture used for program examples

Performance comparisons between program examples are based on measurements taken using the on-chip timer in the transputer. This timer provides a measure of elapsed time in 64-microsecond intervals; thus, the timing measurements will have some variability depending on the state of the timer when the measurements are taken. The performance measurements for each example are generally rounded to a value which is greater than the timer variability. Another source of unpredictable variability in programs with different code sizes is the effect of the on-chip, high-speed memory. Accesses to this memory are much quicker than to the off-chip memory, so that a program with a greater amount of its code on-chip will generally perform better.

occam

The transputer was developed with the specific intention of providing an efficient platform for the execution of the occam programming language. A brief review of the occam programming language is provided here for the benefit of those who may be unfamiliar with the language structure and syntax. This discussion is drawn from the occam 2 Reference Manual†, which provides a complete reference for the language. All of the program examples in this book are written in occam. The word *occam* is a trademark of Inmos Ltd.

Because occam is an indentation-sensitive language, the programs listed in this book use a nonproportional font which maintains an equal spacing of characters, including spaces. Each figure listing is justified on the left margin of the page. Subsequent pages of a single program may be indented from the left margin. This margin indentation preserves the actual program indentation, and the indentations on different pages of a single program are consistent and can be directly compared.

In the program examples, variables are given the name of the value or object they represent. This feature promotes a readable program style. References to the program names within the text are printed using the same font type as is used in the programs themselves so that we may avoid confusion between names from a program with similar words in the text. In-line comments in the code are preceded by a double hyphen.

The most commonly used editor for creating occam programs employs a folding structure which represents portions of a program with a single line beginning with three dots. These folds represent program sections which may be illustrated in other figures. Listing the programs in this way helps to clarify the presentation of many of the larger programs. Each fold is labelled with a comment describing the contents of the fold and, if the program section is illustrated in another figure, labelled with a reference to that figure. See Fig. 3-11 for an example.

The basic program structure in occam is the *process*. A process is an instruction or group of instructions in a program. These instructions initiate a variety of operations, the most primitive of which are assignment and communication. As a simple example of a process consider the assignment of two integer variables, a and b.

```
INT a,b:
a:=b
```

The first statement instantiates, or defines, the variables. Within this first statement, the word INT defines the data type, and the list of variables to be defined follows. An array of variables is defined similarly with an array size prefix of the form [*size*], where *size* indicates the number of elements in the array. The scope

---

†. INMOS Limited, *occam 2 Reference Manual*, (Hemel Hempstead, U.K.: Prentice Hall International (U.K.) Ltd., 1988).

Introduction   *Chapter 0*

of a variable is limited to the process following its definition. The assignment of b to a, done in the second line, is a simple process. Larger processes can be constructed from groups of smaller processes. These smaller processes must then be indented as a group by two spaces.

In describing the temporal relationship between multiple processes, occam is among the most elegant of languages. Any process may execute before, after, at the same time as, or in place of another process. There are three occam statements which define the relationship between multiple processes: SEQ (sequential), PAR (parallel), and ALT (alternate). These three statements, together with the CHAN statement, replicated structures, and control structures, are presented below.

SEQ

The SEQ (sequential) construct causes all of the following processes indented by two spaces to execute in the order listed. For example:

```
INT a,b:
SEQ
  a:=3
  b:=a
```

assigns 3 to a, and then a to b. Together, the four lines of the program can be considered a single process. The SEQ construct is the implicit programming structure found in traditional, single-processor computers.

PAR

The PAR (parallel) statement defines a set of processes which execute in parallel, at the same time. Each individual process included must be indented by two spaces in just the same way as the SEQ process. For example, consider three integer variables, a, b, and c. Using a PAR construct, we can multiply all of these variables by two at the same time:

```
INT a,b,c:
PAR
  a:=a*2
  b:=b*2
  c:=c*2
```

Programs written in occam use a traditional arithmetic structure with the assignment operation performed by :=. The arithmetic operators used are the common symbols (+, -, *, /) with the addition of the backslash (\) for modulo.

The occam programming model does not support shared memory. The following example is not legal, and will not compile, since b is assigned to simultaneously in both processes.

```
INT a,b:
PAR
  b:=a
  b:=2
```

The final value of b cannot be determined, since in a parallel construct it is impossible to predict which process will execute first.

If several processes must use different portions of an array at the same time, the array must be broken down into disjoint subsets of elements using abbreviations. Each of the processes can then uniquely access an abbreviated portion of the original array. For example:

```
[2]INT array:
INT val1 IS array[pointer0]:
INT val2 IS array[pointer1]:
PAR
  val1:=3
  val2:=4
```

The values pointer0 and pointer1 must be defined and assigned earlier in the program. If they are equal, the program will return a run-time error.

Although all of the processes in a PAR construct should, by definition, execute at the same time, on a single-processor computer the parallel processes will in fact have to time-share the cpu.

An additional structure, the PRI PAR (priority parallel) construct, provides a means of executing one process in preference to another. Only when the priority process is unable to proceed further (while waiting on an input or output, for example) can the other processes execute. A PRI PAR structure is written in the same way as a PAR, but with the first process listed in the PRI PAR structure having the higher priority.

The PRI PAR structure is especially useful for programs which need to execute a communication shell at the same time as a normal program task. Typically, the communication should be expedited at the expense of the task, since delaying the communication may mean starving another processor of work. PRI PAR structures are also very useful for real-time systems which must react to external interrupts.

Processes executing at the same time may communicate with each other and can run on physically separate processors. Described in detail in Chap. 2, the PLACED PAR structure configures parallel processes to run on physically distinct processors. This structure essentially associates a complete process with each processor in a network, and defines the link interconnections within the network.

CHAN

Although two parallel processes cannot both assign values to the same variables, they can communicate variables through a CHAN (channel) structure. An input from a channel is performed with a statement of the channel name followed by a question mark and the variable to be assigned. An output on a channel is performed with a channel name followed by an exclamation point and the variable to be communicated. For example, one process can pass an integer value to another through the integer channel `talk`:

```
CHAN OF INT talk:
INT a,b:
PAR
  talk! a
  talk? b
```

The channels are defined in terms of the type of values communicated on them. In the previous example, the channel `talk` communicates integer values (INT). A channel type can also be any of the other variable types (for example, BYTE or REAL32), arrays of such types, or a combination of these. A channel with no defined type is defined as CHAN OF ANY. Channels can also be defined in arrays just as variables are.

The channel communication itself must take place simultaneously in both the input and output processes. This means that the two processes communicating must, at some level, be executing in parallel with each other. If one process wishes to output on a channel and there is no corresponding process doing an input on the same channel, the process attempting to output cannot proceed.

The TIMER channel is a special channel defined in occam. The TIMER definition allows an input from an associated channel to return the current system time. This channel is useful for real-time systems and performance measurements.

```
TIMER time:
INT a:
SEQ
  time? a
```

ALT

The ALT (alternative) construct provides a mechanism for selecting among a group of input processes. In an ALT construct, the first process able to input will proceed and none of the other processes will execute. For example, given two channels, `talk1` and `talk2`, we can write:

```
INT a,b:
CHAN OF INT talk1, talk2:
ALT
  talk1? a
    b:=a*34
  talk2? a
    b:=a+1
```

In this case, if `talk1` inputs a, then a * 34 will be assigned to b. If `talk2` inputs a, then a + 1 is assigned to b.

A processor implements an ALT structure by sequentially testing each of the channel inputs. If several inputs can proceed simultaneously it is not possible to predict which process will be chosen. Just as the PRI PAR provides a means to preferentially execute one of a group of parallel processes, so a PRI ALT will preferentially select one of several simultaneous inputs.

An ALT input can be combined with a boolean guard which will selectively exclude the input if the guard is FALSE. The boolean "guards" the process, either allowing it to, or preventing it from, executing. For example:

```
BOOL go:
CHAN OF INT input1,input2:
INT data:
ALT
  input1? data
    data:=data+1
  go & input2? data
    data:=data+2
```

If `go` is FALSE, `input2` cannot input `data`, even if the channel is available.

There is an ALT structure which is especially useful for real-time systems. When combined with a TIMER channel input, an ALT can be constructed to time out on an input which is delayed.

```
TIMER time:
CHAN input:
INT t,a:
SEQ
  time? t
  ALT
    input? a
      a:=3+a
    time? AFTER (t+1000)
      a:=0
```

Introduction   *Chapter 0*

If `input` does not proceed for 1000 timer cycles after the first `time` input, the second ALT process will perform an input and zero will be assigned to a. The AFTER provides a comparison between the current time and the argument; the `time` input will only proceed after the time of the argument is reached.

Replicated Structures

The SEQ, PAR, and ALT structures can all be replicated, that is, a single statement can define multiple processes.

A replicated SEQ structure is written:

```
SEQ i=start FOR count
```

This statement creates a sequential loop indexed by the integer `i` which is initialized to `start` and repeats `count` times. The parameters `start` and `count` are also integer values. Because the structure is a SEQ, each iteration will proceed sequentially in numerical order. The integer variable `i` is within scope only inside the process and does not have to be defined outside the process.

A replicated PAR structure is written:

```
PAR i=start FOR count
```

This statement creates `count` number of processes which proceed in parallel. Each process is indexed by the integer `i`, whose value ranges from `start` to `start + count - 1`.

A replicated ALT structure requires an array of channels of size `start + count`, and is written:

```
[start+count]CHAN OF INT in:
INT a:
ALT i=start FOR count
   in[i]? a
     a:=a+i
```

This code creates a set of processes, each of which attempts to do an input with its respective element of the array of channels `in`. The first process to do an input on its channel will proceed and add the index value to the input value.

Control Structures

The occam language also includes control structures which permit branching in a program. These structures include WHILE, IF, and CASE statements, as well as subroutine and function calls.

To support the IF and WHILE structures, logical variables are used which can be either TRUE or FALSE. A WHILE statement will repeat a process as long as its associated logical variable is TRUE. For example:

```
INT a:
SEQ
  a:=0
  WHILE (a<4)
    a:=a+1
```

will iterate in the WHILE loop until a = 4.

An IF structure will select the first process in its list whose guard is TRUE. For example:

```
INT a:
SEQ
  a:=0
  IF
    a=4
      a:=8
    a=0
      a:=2
```

will select the second alternative and set a equal to two. Note that at least one of the logical processes must be TRUE or the IF statement will never complete.

IF structures can also be replicated:

```
[start+count]INT a:
IF i=start FOR count
  a[i]=0
    a[i]:=3
```

This process will iteratively test the elements of a. The first element equal to zero will be set to three. If none of the elements is zero, the process will halt. Once again, the values `start` and `count` must be integers.

Logical variables, useful within control structures, can be explicitly defined with the BOOL type. These variables can be used in a logical test and combined with the usual logical operators AND, OR, and NOT. The values TRUE and FALSE can also be used as logical arguments. The following code creates an infinite loop repeatedly setting a to 0:

```
INT a:
BOOL stop:
SEQ
  stop:=FALSE
  WHILE (NOT stop)
    a:=0
```

Logical variables can also be associated with input statements. The following example will only attempt to do an input on channel `in1` if the guard `ok` is TRUE.

```
INT a:
BOOL ok:
CHAN OF INT in1, in2:
ALT
  ok & in1? a
    a:=a+1
  in2? a
    a:=a+2
```

The CASE statement, another control structure, is similar to an IF in that one process from a group is selected and the others are ignored. The CASE statement does not use logical variables, but executes the process whose guard is equal to the argument of the CASE itself. For example:

```
INT a:
SEQ
  a:=3
  CASE a
    7
      a:=a+4
    3
      a:=a/4
    2
      a:=a*4
    ELSE
      a:=0
```

The ELSE process at the end will execute only if no previous processes were executed.

Procedure subroutines are created with the PROC (process) definition. A simple process with two arguments, `double`, is illustrated here. Notice that the type of each argument must be stated.

```
PROC double(INT arg1,arg2)
  SEQ
    arg1:=arg2*2
:
```

A colon indicates the end of a process definition. The procedure is called with a statement of the procedure name:

```
INT a,b:
SEQ
   a:=4
   double(a,b)
```

Functions are defined in a slightly different way. A function is defined with a data type, and must explicitly return a value:

```
INT FUNCTION double(INT arg)
   VALOF
      RESULT (arg*2)
:
```

As with a process, a function is called by using the name of the function with the appropriate arguments:

```
INT a,b:
SEQ
   a:=4
   b:=double(a)
```

In order to assist programmers in the organization and construction of large programs, the occam language supports the use of libraries. These libraries can be defined as separate routines accessible to any program which references the library. A library can be accessed by including the statement:

```
#USE "library_file_name"
```

within the scope of any references to routines or variables defined within the library.

Debugging

Not only do parallel programs inherit all the potential errors of conventional programs, but they also provide new and interesting ways to frustrate users. Since a single node of a parallel computer is, of course, very similar to any conventional computer, all of the errors which one might encounter on a conventional computer are also likely to be found on any of the nodes of a parallel computer.

In addition to these familiar errors, parallel computers provide two new ways to create incorrect programs. These errors are commonly called *deadlock* and *livelock*. Deadlock occurs when a program halts because one of the processors on which the program executes is forced to wait forever while trying to send a message to, or receive a message from, another processor. This condition generally will cause an entire network to come to a halt as each processor tries to communicate with one that is stopped. A very simple example of deadlock occurs when each of two processors simultaneously tries to send a message to the other while neither is listening.

Livelock is a more difficult communication problem, similar to the well-known infinite loop in conventional programs. When a network is livelocked, all of the processors continue to execute some portion of their program, but because of an error in the communication structure of the program, no processor is able to exit the loop. Livelock errors occur much less frequently than deadlock errors, but are also more difficult to trace.

Finding and correcting errors in parallel programs is more challenging than in conventional programs. In a parallel program, many processors may be executing simultaneously and the flow of data communicated between processors can be difficult to track. Tracing a programming error may involve inspecting data in many processors and trying to follow an extremely convoluted control flow.

Despite these difficulties, in recent years utilities have been developed which are very useful in debugging parallel systems. These debuggers typically allow users to easily inspect variables in any processor, check the status of any internal processes within one processor, and examine any data which is being read or output to another processor. These debuggers are most useful for detecting conventional errors or for tracing the control flow of a program which leads to deadlock.

If a debugger is not available, one simple and effective technique for tracing errors in programs is to write a routine which unloads to the host processor some useful parameters from every processor in the network. Using this approach is really equivalent to inserting "print" statements in conventional programs. However, unloading data from every processor has another very important consequence in parallel systems: the successful execution of an unloading routine demonstrates that the processor network is free from both deadlock and livelock at the point in the program where the unloading routine is inserted.

Miscellaneous

Although the programs presented in this book are written exclusively in occam, compilers for other languages do exist. Both C and FORTRAN compilers for programming transputers are available, as well as compilers for some of the less common languages such as Pascal. These languages must have specialized extensions built into them to support the features necessary for parallel programs, such as the ability to create parallel processes and communicate between the processes. The resulting extended languages are less elegant and expressive than occam, but are obviously more familiar to the many programmers accustomed to conventional programming.

The demonstration examples described in the following chapters are simple programs which can execute on a parallel transputer system directly, without the support of any operating system. Operating systems for parallel machines, including transputer-based ones, do exist, however, and can be very useful for constructing programs in a parallel processor environment. These operating systems typically provide to the programmer a communication shell which simplifies interprocessor communication and control among the multiple processors within a network, as well as providing an interface to any other system resources. This shell can also be very helpful when programming errors must be traced. Chap. 6 demonstrates some simple communication programs which perform rudimentary communication functions necessary in a parallel operating system.

In Summary

In this chapter we have considered the basic definition of parallel processor efficiency in computer systems and reviewed the transputer microprocessor itself. The architecture of the parallel processor used to develop the programs we will be considering is based on a toroidal structure. The occam programming language with its unique structures is used for programming all of the examples. Although each of the elements described can be used in a much more complex way than the very simple illustrations given in this chapter, most of the programming examples presented in later chapters use the structures described. For more detailed and complete information, the reader should consult an occam programming manual. The bibliography lists several useful books.

Chapter 1
SISD, SIMD, MIMD, and All That

Computers today come in a bewildering profusion of varieties. One might expect that special-purpose computers would be built in all sorts of shapes and sizes and with varying abilities, but today even general-purpose computers have many different designs. Contemporary computer designers face a host of choices, from the basic transistor implementation to the computer architecture to the operating system software which runs the machine. It is instructive to examine these choices, since various program methodologies useful on transputer systems can resemble many of these computer designs.

Computers are often sorted into architectural classes which describe the connections between a computer's "brain" and its memory. Understanding these classes helps users to appreciate the architectural advantages and disadvantages of the various computers themselves. Within any class, a computer might be constructed using any combination of integrated circuit technologies and be controlled by any suitable operating system software. Although the choice of a particular transistor technology and software may have a great influence on the practicality, usefulness, and performance of a computer, the choice is independent of the architectural class of the machine. Thus one may find architecturally similar computers implemented with very dissimilar technologies.

Computer architecture classifications in use today reflect the history of computing. The first practical computers built used one group of circuits to calculate (the central processing unit, or cpu) and a second group to store instructions, data, and the results of the calculations (the memory). In such computers, the cpu and memory must pass information back and forth (Fig. 1-1) to do any work. No matter how fast or large the cpu or memory becomes, all information passes through this single channel. When circuits were, by today's standards, relatively small, the communication channel could pass more data than the cpu or memory could provide. This is no longer true. Processors today can produce much more data than can be easily moved from one place to another. This information channel between memory and cpu is known as the "von Neumann bottleneck," named for the pioneering computer and information scientist John von Neumann, who first developed many of the concepts used in computer science.

It is interesting to note that as technology has developed through the years, the computing bottleneck has moved from one module to another. Originally, the cpu was the most difficult device to build and the cpu itself constrained the applications for which computers were appropriate. As integrated circuit technology de-

SISD, SIMD, MIMD, and All That   Chapter 1

Figure 1-1
An SISD computer with three parts: a central processing unit (cpu),
a memory, and a communication channel through which information passes

veloped, the power of cpus grew, and the amount of memory within a computer limited its performance. Today, memories are much larger and faster and cpus are more powerful than ever before, and we have reached the point where the von Neumann communication bottleneck is often the most significant problem facing a designer of high-performance computers.

In order to conquer or avoid this communication bottleneck, designers have developed alternatives to the simple architecture shown in Fig. 1-1. The simplest approach to improving the performance of a computer which is limited by its communication bandwidth is to increase the size of the communication channel. Doing this may not actually solve the bandwidth problem, but it at least defers the problem, although at the cost of larger and more expensive systems. Unfortunately, this cost is becoming increasingly less affordable and designers can no longer simply increase the size of a communication channel to improve its performance. The pins devoted to communications on computer chips are often more than half the total number, and increasing their number further is dauntingly expensive. Nor can signals be moved more quickly through channels since the speed of light appears to be an ultimate limit and the propagation of signals in today's fastest computers is already facing that limit.

Because of the cost of bigger or faster channels, an alternative to using a single channel must be found. The most natural alternative, multiple channels, requires either (or both) more memory modules or more cpu modules. This is the approach that nearly all of today's high-performance computers are taking. Since there is no limit, other than financial, to the number of processors or memory modules a computer performance enthusiast might desire, and there is no limit to the number of ways to interconnect these modules, researchers have had a field day designing and publishing reports of new computers, both theorized and realized.

Because there exists such a plethora of ideas, it is important that we classify the various computing machines so that they can be sensibly compared and evaluated. In the most common classification method, a machine is described in terms of its data and instruction streams, that is, whether it has one or many independent processing units each with its own instruction stream, and one or many communication channels, each providing an independent data stream. These classes are somewhat imprecise and can be interpreted in different ways depending on one's perspective, but they are still helpful and instructive.† Thus the computer shown in Fig. 1-1 is called a single-instruction (because there is only one cpu which executes one instruction at a time), single-data (because there is only one data store and one channel by which data is retrieved) computer, or an SISD computer.

Multiple Data Paths

It is a simple matter, given our basic three computer building blocks, to design a new parallel computer. The difficulty is in creating a design that is efficient, flexible, cost-effective, and intelligibly programmable. Perhaps the simplest approach to parallelism is to perform identical operations simultaneously on many pieces of data. A computer that does this avoids the traditional von Neumann bottleneck by storing data in many independent devices, each connected separately to a processor. This approach, known as single-instruction, multiple-data, or SIMD, is the usual approach taken to implement very large parallel computers. This technique is popular with designers of high-performance computers because it simplifies the programming task; only one program must be written regardless of how many data elements are operated on at once. On the other hand, one *must* operate on many pieces of data in the same way simultaneously or the approach is a waste of time. Fortunately, many difficult problems in science and engineering do require such manipulation of massive amounts of data.

Because it is difficult to build many independent cpus, each cpu in a parallel computer is usually quite simple and uses identical circuitry (Fig. 1-2). Each circuit module is a complete processor, but without independent program control, so that

---

†. This taxonomy was originally developed by M.J. Flynn in "Some Computer Organizations and their Effectiveness," *IEEE Transactions on Computers*, C-21, No. 9 (Sept. 1972), pp. 948-60.

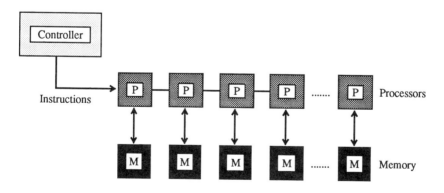

Figure 1-2
A single-instruction, multiple-data computer executing one instruction at a time using many simple processors to process many data elements at once

every processor executes the same instruction at the same time. Modern SIMD machines generally also have the ability to selectively disable operations on particular processes depending on local data values.

Since each of the small cpu elements is very simple and must execute the same instruction as all of the others at a given time, these small units can be made and controlled in vast quantity. Current machines have tens of thousands of processors and can demonstrate extraordinary performance. However, nothing in life is free. The processors in an SIMD computer usually have very limited ability to communicate with each other and the controller. If *communication* between the processors or the host is important for a particular application, the performance of an SIMD computer will suffer badly.

Among today's conventional supercomputers are SIMD machines with a somewhat different flavor. These supercomputers are generally termed "vector" machines because they operate on an entire vector or array of data at once. The circuitry which performs operations in parallel, however, is not truly a multiprocessor. The parallel operations in a vector supercomputer tend to be much more complex than the parallel operation on most SIMD parallel computers, but operate on much smaller data sets, generally 128 elements or fewer. Indeed, some observers do not classify vector supercomputers as SIMD machines, despite their specialized hardware. These conventional supercomputers are reaching performance plateaus, especially because of signal propagation delays. Because of such limitations, supercomputer companies are also beginning to incorporate some of the multiple processor techniques discussed below.

Multiple Instruction Paths

Shared Memory

The natural opposite to a single-instruction, multiple-data machine is a multiple-instruction, single-data computer. Strictly speaking we can see that no such computers are actually built, since each of the multiple instructions requires separate hardware, and this hardware also stores data objects, providing multiple data paths. However, computer systems with restricted access to a general storage device do exist, and are commonly called *shared-memory system*s. Such systems will typically have multiple processors connected to a single storage device through a common communication channel, usually a standard computer bus of some sort. Figure 1-3 shows a logical representation of such a shared-memory system. Notice that, in addition to the primary storage, every processor also has local memory storage with independent access to the storage. Therefore, a practical shared-memory computer is a multiple-instruction, multiple-data (MIMD) computer with a common interprocessor communication channel. Parallel computers of this sort clearly need a very high-performance communication channel to accommodate the many processors moving information around. It is also clear that shared-memory systems cannot expand themselves endlessly since the channel bandwidth is constant

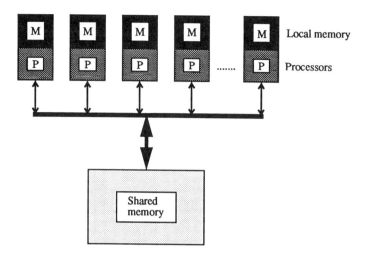

Figure 1-3
A practical shared-memory computer with a small local memory
for data cache and program storage as well as a common memory

and every processor must work through it. This lack of expansibility is in sharp contrast to SIMD machines of the type shown in Fig. 1-2.

Shared-memory computer systems may require complex data control methods to ensure the integrity of the data used in a program. Since every processor can access all of the data, any one processor can destroy or misuse information allocated to another. To prevent this, information in the shared memory is generally "locked" by one processor when in use and cannot be accessed by another. When the first processor has finished working with the information it is "unlocked" and made available to the next user.

A shared-memory system is the first parallel system described so far that has the capability of actually running more than one totally independent program at a time. A single-instruction machine has a sequential control flow, even though it may be doing a particular operation in many processors and on many pieces of data at the same time. Since a shared-memory system can execute entirely different programs at the same time, the control flow is much more complex and obscure. Processors can now easily interact with each other rather than being limited to interacting primarily with data.

Effectively controlling multiple processors is a difficult task. Most people are accustomed to one-track programming and find using multiple-track programming problematic. Moreover, control is not the only thorny issue. Parallel algorithms can be difficult to devise and implement efficiently. An SIMD machine uses a fixed approach to solving a problem in parallel: identical operations must affect multiple data elements concurrently. A shared-memory system is not limited to using this approach, and problems can be divided up in many different ways to make use of the multiple processors available. Some of these approaches, suitable for implementation on transputer systems, are described in the chapters that follow.

Processors in a shared-memory machine generally interact in one or both of two ways: through direct message-passing from one processor to another, or by leaving information in the common memory. Message-passing is the more direct approach but also involves more hardware, since the communication channel must be able to handle messages and the processors must be able to signal each other in some way through interrupts, message buffers, or the like.

Using the common memory as a repository for messages can provide versatility. Processors can scan through variously allocated blocks of memory leaving messages, receiving messages and instructions, and generally treating the common memory as a combination postal service and safety deposit box. Indeed, the programming techniques used to drive a shared-memory machine often make it look a little like a data-flow computer, with the memory controlling the processors rather than the other way around.

Real implementations of shared-memory machines generally use traditional, von Neumann, SISD processors as the actual cpu engines. The local memory at each processor is used to store programs and cache data taken from the shared memory. This approach reduces the bandwidth requirements of the communication channel, but requires that every processor hold a copy of the program. If each processor runs a different program, no memory is wasted. If each processor is essentially running the same program, however, the identical program will be stored in each processor, a redundant waste. The inverse relationship between the amount of program memory and the bandwidth required to move program code is often demonstrated in parallel computer design.

Distributed Memory

If a multiple-instruction, multiple-data computer does not include a shared memory, it is considered to be a *distributed-memory system*. Distributed-memory MIMD computers can use either a single communication channel for interprocessor communication, as do shared-memory computers, or multiple communication channels. Transputer-based MIMD computers use the latter technique, since the multiple transputer links form point-to-point processor communication channels. A distinguishing feature of distributed-memory MIMD machines is that one processor cannot access the memory of another. In fact, for multiple-cpu computers, the distinction between shared and distributed memory is much more useful than the distinction between multiple and single data streams. Figure 1-4 is a simple diagram of a generic MIMD distributed-memory parallel computer.

In some ways, a distributed-memory MIMD computer is the most complex design we will discuss. It has the complexity inherent in both the single-instruction multiple-data approach and the shared-memory approach. An SIMD machine must distribute its data in a useful way; a shared-memory machine must distribute its program in some way; and a distributed-memory MIMD computer must do both.

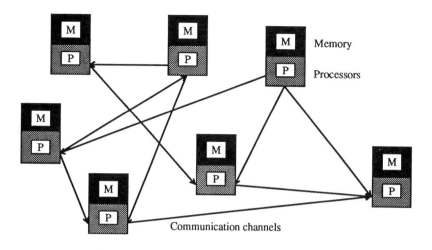

Figure 1-4
A distributed-memory MIMD computer with many communicating processors, each with its own local memory but no shared memory

The advantage of this complexity is the resulting performance. A shared-memory MIMD computer inevitably faces limitations in its ability to access data from a common memory, while a SIMD machine faces interprocessor communication difficulties. In contrast, a distributed-memory MIMD computer circumvents both of these limitations, but at the expense of programming complexity.

MIMD computer hardware ranges from very simple, replicated processor units to very complex, high-powered cpus composed of many integrated circuits. Typically, the more integrated circuits there are in a processor node, the fewer nodes there are in the computer. This circumstance points out a basic trade-off available to parallel computers which is not found in typical SISD machines. The speed of an SISD computer can only be increased through hardware improvements, while the speed of a parallel computer can be increased either by the addition of processor nodes or by the improvement of the hardware at each node. Generally it is more cost-effective to add processors; however, a thorough analysis of this trade-off for a particular parallel computer can be quite difficult.

In common with the SIMD organization, MIMD machines can grow fairly readily in size and performance. Because the number of data paths and amount of memory usually increase with the number of processors, there is no obvious limit to the number of processors that can be used and the performance an MIMD computer can realize. Limitations in performance do exist but the limits are reached in more subtle ways, generally because of computational inefficiencies, communication requirements, or hardware constraints.

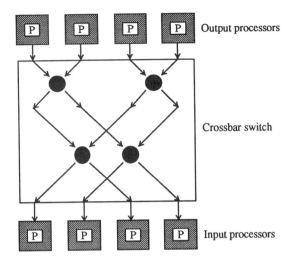

Figure 1-5
An MIMD computer with a crossbar switch connecting every processor to every other with the switches represented by ●

Distributed-memory computers can be further distinguished from each other by the flexibility of their interconnections. The interconnection in a system can either be switched or not switched. If the processor interconnects are not switched, they are either point-to-point, that is from only one processor to only one other processor, or broadcast, from possibly any one processor to all other processors. The most direct implementation of a transputer system uses point-to-point interconnections between transputer processors since the transputer links themselves form point-to-point interconnections.

Switched-interconnect systems also use point-to-point interconnections, but those interconnections can be changed under program control. In such systems, each processor's links are connected to a switching network (often a crossbar switch) as shown in Fig. 1-5. This switching network provides the ultimate flexibility in real-time reconfigurability but is expensive and can be problematic to control. The same effect can be achieved through software in a point-to-point system although with much lower performance. The issue of switched versus non-switched interconnection can be reduced to a choice between increased hardware expense and correspondingly higher performance on the one hand, and software cost and communication overhead on the other hand.

In addition to cost and control, the limited expansibility of a switched system can be a problem. The number of switches needed to completely connect processors together grows faster than does the number of processors. In a large system, completely interconnecting all of the processors is both expensive and difficult. Sometimes multilayered switching systems are used to reduce the total number of switches needed. Using this technique adds delay in switching but reduces the total number of switches needed. Regardless of the switching scheme used, switching is expensive.

Architectural Issues

Any computer faces inherent limitations in its ability to do work. The original single-instruction, single-data machines were limited by the speed of the cpu and the rate at which it could get information or work to do. The same limitations are of course equally applicable to parallel systems, but in such systems the bounds are somewhat more elastic. The performance of old-fashioned SISD computers was increased by making the cpu faster or the communication channel larger, and these same improvements can be applied to parallel machines.

We can also speed up a parallel machine by increasing the number of cpus and communication channels. Since we can presumably increase the number of channels and processors forever, if we are rich enough, what performance limits will parallel machines ultimately face? One limit mentioned earlier is the difficulty of getting information into and out of the machines themselves, but channels to do this can be replicated, and storage and display facilities can be built in parallel as well.

In addition to practical constraints on the performance of parallel computers, there are natural performance limits internal to the parallel machines themselves. One such limit is the overhead involved in distributing work. If more than one processor is working on a task, the processors must communicate, if only to agree on when they are finished. If each processor has a different task and none of the tasks require communication, there is no limit to the performance of the computer; however, in this case the processors are really separate machines and can hardly be thought of as working "in parallel."

One fundamental performance limitation of parallel computers, then, is the cost of communication. The more processors and communication channels there are in a computer, the higher this cost, since it will either cost more to switch the communication channels or take longer to pass a message through more processors. A system with instantaneous point-to-point communications connecting all of the processors would help but has yet to be built. In any case, communication overhead in a multiprocessor computer does not decrease when processors are added as does the work per processor, but rather increases or, at best, stays the same.

A second fundamental limitation to the performance of parallel computers is related to the type of problem being solved. Although there is some controversy about this issue, it appears that certain problems, perhaps the majority, cannot be solved in parallel; their mathematical structure will not permit it. Fortunately there are many useful problems which *can* be successfully solved in concurrent pieces. Even so, the algorithms which permit us to solve these problems often involve much communication. In an effort to reduce the total amount of communication ordained by a particular algorithm, computer scientists devise many special-purpose architectures, matching the architecture to the algorithm.

Given the fact that parallel computers are practical for solving certain problems, which designs are best? Does the choice of design make any difference? In a very general sense, all computers *are* equivalent. What can be done by one can be done by all; only the performance efficiency changes. This is obviously true if one considers that a sequential, single-instruction computer can do the same tasks as a parallel machine, merely doing the tasks one at a time. It is not so clear that the parallel MIMD and SIMD machines are equivalent. An MIMD machine can be programmed so that each processor acts identically just as each processor of an SIMD machine does; but what about the reverse? Essentially, the SIMD machine can do the same tasks as an MIMD, but must do them sequentially.

A simple example of this is a CASE or IF statement in which, depending on a local value, a processor chooses an operation. Since each processor in an MIMD machine is running an independent program, the processors together can execute different instructions simultaneously. In contrast, an SIMD machine can only execute one instruction at a time. This means that each instruction path must be run sequentially, one operation after another. In the extreme case, if every element of an array required a different computation, an SIMD machine would be reduced to sequentially processing one element at time.

Distributed-memory and shared-memory systems can also be considered equivalent. It is quite easy to see how a shared-memory computer can behave like a distributed-memory computer. If a shared-memory computer divides its memory so that each processor has a partition, and if the shared-memory computer is programmed so that each processor accesses only its own partition, it is essentially behaving like a distributed-memory computer. Some mechanism must be included to accommodate interprocessor communication, but this is easily done by allowing shared blocks of memory to be accessed only by those processors which must communicate. This arrangement can accommodate broadcasts, in which a piece of data is sent to every processor, by having all processors share one small block, or point-to-point communication, in which only pairs of processors can share a message block.

It is considerably more difficult and less efficient for a distributed-memory machine to mimic a shared-memory machine. It can be done, however, by distributing the "shared" memory over the array as is done with any distributed computer. A memory handler is placed in every processor to intercept requests for data. If the data requested are not stored locally, messages are sent to the processor with the needed information. The data are then passed back to the requesting processor. When all of the processors are actively requesting data, the communication load can be quite heavy. The same precautions regarding data integrity must be taken as were mentioned earlier for shared-memory systems.

It is also possible to replicate a "shared" memory at each processor. If every processor has a copy of the data, communication overhead is greatly reduced. As usually happens in this kind of arrangement, however, the communication overhead is reduced at the expense of using n times more storage for n processors. In any case, communication overhead is not eliminated since after each operation the data at each processor will have to be communicated to every other processor, in order that the integrity of the shared data be maintained.

Classifying a particular parallel computer can be quite difficult. As we have seen, one type of computer sometimes mimics the behavior of other types. Although classifications are usually based on hardware features, particular machines often have features from different classes, making them hard to classify. For example, the multiple SIMD (MSIMD) computer can use many (or at least several) SIMD machines together. An MIMD computer with a vector processor at each node could be considered an MSIMD computer, as could some of the new supercomputers being built with more than one processor.

The distinction between shared-memory systems and distributed-memory systems is also blurred if a distributed machine uses a direct broadcast facility. The broadcast acts as a shared facility which can be accessed by all processors simultaneously.

In Summary

In this chapter we have considered a variety of parallel processing architectures. Parallel processing computers are generally classified according to the number of different instructions which can be executed simultaneously and according to the number of communication channels used to access data. In addition, parallel computers are often classified as shared-memory (if all of the processors can access a given memory location) or distributed-memory (if only one processor can access a given memory location). Parallel computers can be built with various features from different categories, making them difficult to classify.

The performance of parallel computers can be expressed in terms of number of processors, amount of memory, and communication bandwidth. Each variety of parallel computer will have its particular limitations in ultimate performance. Shared-memory machines generally encounter a hardware limit to their performance, usually due to contention between processors for the shared memory resources. Distributed-memory computers have no such contention but encounter overhead from communication requirements. This overhead tends to depend on the task performed and is difficult to quantify, but will eventually impose an ultimate limit on performance.

Transputer systems are most often used in distributed-memory multiple-instruction, multiple-data parallel computers. The transputer's point-to-point links offer communication channels which are independent of any memory-sharing requirements. Arbitrarily large systems can be built using these processors. Local memory provided on each processor chip, hardware support for task switching, and elegant software support for distributed parallel processing make the transputer especially appropriate for constructing distributed-memory MIMD computers.

Such computers have a wide variety of uses and can be programmed in many ways. These advantages make the transputer a highly flexible tool. Techniques used to achieve this flexibility are presented in subsequent chapters.

Chapter 2

Architectures

One of the most popular architectural classes of parallel computers is the multiple-instruction, multiple data (MIMD) class. Most transputer-based computing machines fall into this class. Although members of a single class of computing machines, multiple-instruction, multiple-data computers are extraordinarily diverse in their architecture. Even though every node in an MIMD computer may be identical, the number of ways in which the nodes can be connected is very large, and every different architectural or interconnect scheme gives rise to a different computer with its own special abilities and limitations.

Any MIMD architecture can be classed as either regular or irregular. Most MIMD computers use a regular architecture, one that can be described with a mathematical formula. In contrast, an irregular architecture cannot be described with a formula and therefore must be described with a list of the nodes in the network and their communication channels. However, any subset of an irregular network may be regular, or a network may be regular except for a few aberrations which make it irregular.

A regular architecture can generally be scaled proportionately without its essential structure changing. But just because a network is scalable does not necessarily mean that each processor node in the network will remain the same as the network is scaled. Indeed, some network architectures actually require that processor nodes add communication channels as the network grows. Other networks can only grow by a fixed amount at a time.

Regular networks have some distinct advantages over irregular networks. One significant advantage is that a regular network can be changed in size without its fundamental characteristics being altered. Thus, a parallel computer built around a particular regular network can have different numbers of nodes, making it large or small, fast or slow. A second advantage is that regular networks are well defined so that programs written for them can be somewhat portable between networks of different sizes and between computers with the same architecture. Thus a properly written program for a regular network can also execute on a larger or smaller network with the same architecture. Likewise, a program written for one parallel computer may be portable to another computer with the same architecture. Other advantages are that regular networks are more modular, easier to construct, and often easier to program than irregular networks.

In contrast to the generality of regular networks, irregular network architectures are usually made to match specific problems. Such networks can be used for general computing but are more often found in special-purpose computers. Parallel computers with irregular architectures tend to be built with nodes of different capabilities at different locations in the network. Even if the same cpu is used for each node, the nodes often have different supporting hardware. These peculiarities in connection and capability within a network can make programming problematic. The various methods used for implementing parallelism demonstrated in this book cannot automatically be used in an irregular network without the special resources and connections which might be available in such a network being wasted. On the other hand, the parallel implementation of a problem on an irregular network is generally obvious, since the network is usually chosen to suit the problem very closely. What is often not so obvious is how to organize the problem and design a suitable network in the first place.

Irregular networks are used most often with pipeline parallelism, and regular networks with distributed data parallelism. However, either mode of parallelism can be used with either type of network. Indeed, pipeline parallelism can easily be implemented on a regular network in order to build a system very similar to a systolic processor. These issues will be discussed in more detail in later chapters.

Transputers are ideally suited to the construction of a wide variety of MIMD network architectures, both regular and irregular. Because the transputer's links provide such a simple mechanism for processor interconnection, it is easy to build or modify any particular network. And because the occam language provides a highly flexible programming tool for parallel systems, programming the many possible networks is not especially difficult.

There is, however, one major restriction on the construction of networks with transputers. Because the number of links in each processor is fixed, transputers cannot be used to construct networks with more communication channels than there are links, unless link-switching devices are used. These switches can provide greater interconnectivity in a network but do so at the cost of reduced performance and increased expense and complexity.

Configuration Descriptions

Transputer networks programmed in occam are defined by a configuration description. Clearly, this description must match the physical network, or a program meant to run on the network will not execute properly. If a program uses only a portion of a network, only that portion must be defined in the configuration description. This configuration description must include each processor in the network, the type of each processor, and the link interconnections between the processors. In addition, the configuration description must associate each processor with an individual program which that processor executes. The program must also match the physical network if it is to execute.

In occam, a channel is a programming abstraction through which two parallel processes can communicate in one direction; where appropriate, a channel is associated with a hardware link on a processor. The processor links physically inter-

Chapter 2 — Architectures

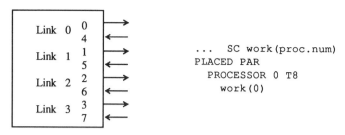

Figure 2-1
Transputer link and address assignments

Figure 2-2
A single-node network configuration description

connect a network of processors. Thus one complete interconnection between two processors is defined in a configuration description with two channels, one for the input from a processor, and one for the output to the same processor. The processor used in the following examples is a T800 transputer with eight unidirectional links, so that each processor can be associated with eight channels. Processors that become available in the future may have more links which are arranged in different ways.

Each channel is associated with a particular link on a processor by means of a PLACE statement matching the channel with that link's hardware address. Since the links are actually used in pairs (for input and output), two channels are used to completely connect a link on one processor to a link on another. The link pairs have the addresses 0 and 4, 1 and 5, 2 and 6, and 3 and 7. The link pairs are conventionally considered to be a single bidirectional link and named link 0 through link 3 (Fig. 2-1). Addresses 0 through 3 are all output links, and addresses 4 through 7 are input links.

The simplest "network" that can be constructed is a single processor with no interconnections. The configuration description for such a network is shown in Fig. 2-2. The first line indicates a fold representing the actual program being run on the processor. The SC indicates that the fold is an independent program unit (separately compiled). A copy of the first line of the program is included on the fold line to demonstrate any formal parameter names used in the program. The values of the dummy parameters are passed from the configuration description. In this simple case, the processor number, a 0, is passed to the program from the configuration description and assigned to the parameter proc.num.

The configuration description continues with a PLACED PAR construct on the second line. This construct is similar to a PAR as used in occam but, in this case, places the construct on a physical processor. Including a PAR is necessary because most networks have more than one processor, all of which must run in parallel. The processor is defined on the third line as a T8 processor and given the arbitrary number of 0. Each processor in the network must have a unique processor

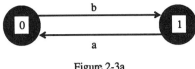

Figure 2-3a
A two-node network

number to distinguish itself from the other processors in the network. The fourth and final line actually associates the program work with processor 0.

Since there is only one processor in the "network," this very simple configuration description lacks any link connections to other processors. A more complex, two-processor system with one link is shown in Fig. 2-3a. Figure 2-3a shows two processors, arbitrarily called processor 0 and processor 1, which communicate over two channels, a and b. Processor 0 talks to processor 1 through channel b, while processor 1 talks to 0 through channel a. Recall that each physical hardware link on the transputer is actually either an input or an output link. A pair of channels, one for input and one for output, is generally considered to be one complete, bidirectional link, so an interconnection is usually drawn with one line. Except for Fig. 2-3a, every drawing in this book uses this one line representation for a bidirectional link. Nonetheless, the configuration description does require that both channels be explicitly defined.

The configuration description for the two-node network (Fig. 2-3a) is shown in Fig. 2-3b. Within the PLACED PAR structure, two processors are defined. Each processor runs the same program, work, and uses the same dummy processor numbers as in Fig. 2-3a. The two processors communicate over a pair of channels named a and b. Processor 0 uses channel a as an input which is placed at address 4. Channel b is placed at address 0 as an output. Notice that these links are associated with the in and out channel variables in the program work. In this way,

```
...  SC work(CHAN OF ANY in,out)
CHAN OF ANY a,b:
PLACED PAR
  PROCESSOR 0 T8
    PLACE b AT 0:
    PLACE a AT 4:
    work(a,b)
  PROCESSOR 1 T8
    PLACE a AT 0:
    PLACE b AT 4:
    work(b,a)
```

Figure 2-3b
Configuration description for a two-node network

any output to channel out in processor 0's program will be output on link 0 over channel b.

Processor 1 uses the same configuration structure as processor 0 except that the channel names must be reversed. Whatever is output from processor 0 is input to processor 1. Thus processor 1 will use channel a as an output and b as an input. Notice that the addresses in the PLACE commands are reversed and that the argument order in work is reversed. This exchange of input and output channels must be consistently programmed for every link connecting a pair of processors in any network.

Since both processor 0 and processor 1 are using link 0 as their communication channel, the configuration describes two processors, each of which has link 0 connected to the other processor. Any link could just as well have been used by either processor.

Once a transputer network is configured, it must be booted. A transputer can be booted either by link or from memory, but most often transputer networks are booted through the network links. In order for this to happen, at least one processor somewhere in the network must have a free link, one which is not connected to another processor in the network. This free link is unconnected according to the configuration description, but must actually be connected to a root processor which passes the initial booting program down the free link. The processor with the free link must be listed first in the configuration description so that the compiler knows where the root node is connected in the network, and so that the compiler can create the correct boot code for the network.

If, in a network, every processor's links are all connected to other processors, one of the processors in the network must be booted from memory. Since the two-node network in Fig. 2-3a connects only one link, any one of the other links may be connected to a root processor. If, instead of having only one link from each processor connected, all of the links were actually connected in a closed network, no link would be free for booting and one of the processors would have to boot from memory.

In the remainder of this chapter, the configuration of an irregular network and a variety of interesting or popular regular networks is described, and a few comments about their implementation with transputers are made. The irregular network is configured as a more complete example than the two-node network shown in Figs. 2-3a and 2-3b.

Networks

Irregular Networks

An example of an irregular network is shown in Fig. 2-4a. This network is not developed for any particular application but is simply used for illustration and is also found in Chap. 7. The corresponding configuration description is listed in Fig. 2-4b. Notice that for an irregular structure such as this one, every processor

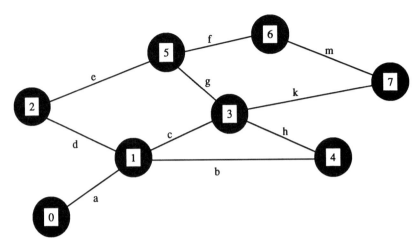

Figure 2-4a
An irregular network

```
...   SC work4(CHAN OF ANY in0,out0,in1,out1,in2,out2,in3,out3)
...   SC work3(CHAN OF ANY in0,out0,in1,out1,in2,out2)
...   SC work2(CHAN OF ANY in0,out0,in1,out1)
...   SC work1(CHAN OF ANY in.0,out.0)
[2]CHAN OF ANY a,b,c,d,e,f,g,h,k,m:
PLACED PAR
  PROCESSOR 0 T8
    PLACE a[1] AT 0:
    PLACE a[0] AT 4:
    work1(a[0],a[1])
  PROCESSOR 1 T8
    PLACE a[0] AT 0:
    PLACE a[1] AT 4:
    PLACE b[0] AT 5:
    PLACE b[1] AT 1:
    PLACE c[0] AT 6:
    PLACE c[1] AT 2:
    PLACE d[0] AT 7:
    PLACE d[1] AT 3:
    work4(a[1],a[0],b[0],b[1],c[0],c[1],d[0],d[1])
```

Figure 2-4b
Configuration description for an irregular network

```
PROCESSOR 2 T8
  PLACE e[0] AT 6:
  PLACE e[1] AT 2:
  PLACE d[0] AT 3:
  PLACE d[1] AT 7:
  work2(e[0],e[1],d[0],d[1])
PROCESSOR 3 T8
  PLACE c[0] AT 0:
  PLACE c[1] AT 4:
  PLACE h[0] AT 5:
  PLACE h[1] AT 1:
  PLACE k[0] AT 6:
  PLACE k[1] AT 2:
  PLACE g[0] AT 7:
  PLACE g[1] AT 3:
  work4(c[1],c[0],h[0],h[1],k[0],k[1],g[0],g[1])
PROCESSOR 4 T8
  PLACE h[0] AT 2:
  PLACE h[1] AT 6:
  PLACE b[0] AT 3:
  PLACE b[1] AT 7:
  work2(h[1],h[0],b[1],b[0])
PROCESSOR 5 T8
  PLACE e[0] AT 1:
  PLACE e[1] AT 5:
  PLACE g[0] AT 2:
  PLACE g[1] AT 6:
  PLACE f[0] AT 7:
  PLACE f[1] AT 3:
  work3(e[1],e[0],g[1],g[0],f[0],f[1])
PROCESSOR 6 T8
  PLACE f[0] AT 2:
  PLACE f[1] AT 6:
  PLACE m[0] AT 7:
  PLACE m[1] AT 3:
  work2(f[1],f[0],m[0],m[1])
PROCESSOR 7 T8
  PLACE m[0] AT 0:
  PLACE m[1] AT 4:
  PLACE k[0] AT 1:
  PLACE k[1] AT 5:
  work2(m[1],m[0],k[1],k[0])
```

Figure 2-4b (cont.)
Configuration description for an irregular network

must be explicitly listed together with each connecting channel. All of the processors with the same number of link connections run the same program: `work4` for processors with four connected links, `work3` for those with three connections, and so on.

Each bidirectional channel in the example is organized as an array of two elements. This arrangement makes it easy to perform the input/output exchange of links necessary for proper communication. For each connected pair of processors, the zero element of a channel is the input for one processor and the output for the second. The one element is the input for the second processor and the output for the first.

Regular networks are much less tedious and more interesting to define than are irregular ones. PLACED PAR structures can be replicated just as PAR structures are in occam, and, with a judicious use of these replicated structures and careful organization of channels, large, regular networks can be easily defined.

Rings

Among the classes of regular networks, a ring of processors is the simplest. Each processor in a ring requires a bidirectional link for each of its two neighbors. Figure 2-5a shows a simple four-processor ring. In Fig. 2-5b is listed the configuration description for this network. The first two lines of the description represent the program running on every processor. To each processor are passed its processor number, p, and four channels, the input and output channels on both the left and right sides of each processor.

Configuration descriptions can include simple parameters which assist in the definition of the network. For the ring network shown, the parameter `num.trans` is defined to be four, the number of processors in the network. Channels a and b are defined as an array of size `num.trans`. For this network, a and b are used to exchange the input and output channel pairs for connected processors, and the array elements define the connection from processor to processor.

The ring network shown in Fig. 2-5a is configured using only one PLACED PAR structure. The configuration uses two pairs of channels, a and b, which are associated with each processor, creating bidirectional link connections to the processors on both the left and the right. For the nth processor, the channel on the left is the nth channel element for both channels a and b (input and output) and the channel on the right is the nth+1 element for a and b. Thus processor 0 is connected on the left to channel element 0 and on the right to channel element 1. The rightmost processor, however, is a special case and must be connected to channel element 0 on the right. To correctly connect this channel, we define the value q as the processor number plus one modulo the ring size. Thus, value q will be the right-side channel element number for every processor including the right-most one, for which q will be 0. The processors are labelled and the channel element numbers are shown in Fig. 2-5a near the links they represent.

In this ring network, each processor is physically connected to hardware link 0 on the left and hardware link 2 on the right. To maintain the correct input and

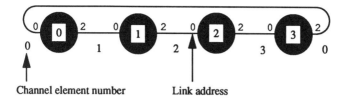

Figure 2-5a
A ring network

output connections, the right-hand link channels are reversed with respect to the channels on the left, both in the program definition and the address assignment. The program call then correctly associates the links with the channels `left.in`, `left.out`, `right.in`, and `right.out` in the program itself. Figure 2-5a shows the link addresses for each processor and its link connections. Because of its structure, we can extend this configuration of processors to a network of any size simply by changing the `num.trans` definition.

Ring networks are easy to program and use, and can be infinitely scaled to larger and larger networks. No matter how large the network becomes, each processor requires only two bidirectional links. Thus a single hardware module can be used repeatedly to build ever-larger parallel computers. The network can also grow one processor at a time while maintaining its ring structure, allowing for very fine control of the number of processors in a network. In addition, this type of network is simple to program for both pipeline and distributed data applications.

Simplicity, ease of construction, and scalability are a ring network's greatest advantages. But a ring network also has a great drawback. The largest interprocessor distance between any two processors in a ring is one half the number of processors in the ring. This distance creates significant overhead when one processor

```
... SC work(VAL INT p,
       CHAN OF ANY left.in,left.out,right.in,right.out)
VAL num.trans IS 4:                       --ring size
[num.trans]CHAN OF ANY a,b:
PLACED PAR p=0 FOR num.trans
  VAL q IS (p+1)\num.trans:               --next channel
  PROCESSOR p T8
    PLACE a[p] AT 4:                      --left
    PLACE b[p] AT 0:
    PLACE b[q] AT 6:                      --right
    PLACE a[q] AT 2:
    work(p,a[p],b[p],b[q],a[q])
```

Figure 2-5b
Configuration description for a ring network

communicates with another. As the network grows, this distance and the associated communication overhead grow at the same rate.

Because of their communication overhead, rings are most useful with pipeline parallelism and distributed data parallelism requiring little communication. One-dimensional problems are ideally suited to the one-dimensional structure of a ring. Higher-dimensional problems can be implemented by distributing the data in only one dimension.

Toroids

The next logical step up in complexity from a ring network is a toroid, a two-dimensional array of processors with the sides wrapped around. Figure 2-6a shows a toroidally connected network of sixteen processors. Each processor has four others connected to it, one each on the left, right, top, and bottom. Figure 2-6b shows the configuration description for this toroid.

The configuration description begins with two parameter statements which define the size of the network as x.trans by y.trans for the x and y dimensions respectively. Following the network dimension statements, the code for each processor is defined in the separately compiled fold. Each processor receives parameters defining its position in the array (i, j) and the size of the entire array (x, y), followed by the eight one-way channels for the left, right, up, and down connections.

After the process declaration, the external channel array is defined in one large array. Since there is one channel for each connection between pairs of processors, there is exactly one channel for each processor in the horizontal and one in the vertical dimension. The channel definition thus creates an array of x.trans by y.trans channels with two additional dimensions, one for the horizontal and vertical set of channels, and a second for the input and output pairs necessary to implement a bidirectional link.

The processor configuration is created with one doubly nested PLACED PAR structure. The structure is replicated in two dimensions, x and y, using the i and j variables. The processors are ordered from upper left to lower right using the value p. As with the ring configuration, each processor is connected on the left to the channel whose element is the same as its processor number and on the right to the channel element one greater. The vertical assignment is similar. Since one array of channels is used to connect the processors in both the horizontal and vertical directions, the horizontal and vertical channels are distinguished by subscripting the channel array using the horz and vert constants. The a and b elements perform the input/output exchange, just as they did with the ring.

A toroid wraps its link connections around in both the x and y dimensions. This link connection is accomplished in the same way that the ring created a closed processor loop. Two values, m and n, are defined to be one greater than the i or j replicator modulo their respective dimension size. Thus value m will be the right-side channel element for every processor including the right-most ones, while n

Chapter 2 Architectures

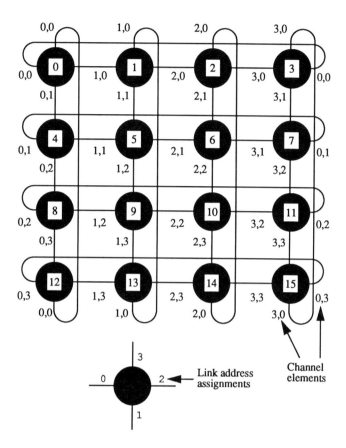

Figure 2-6a
A toroidally connected network

will be the bottom-side channel element for every processor including those on the bottom row.

In the toroidally connected network shown in Fig. 2-6a, each of the processors and the link connections are labelled as they are defined in the configuration description. Each processor uses link 0 to connect to the processor on the left, link 2 on the right, link 3 above, and link 1 below.

Notice that a toroid is a closed surface. If each processor in a toroid has only four links, none is available to boot the network. In this case, either one processor in the network must be booted from memory, or the toroid must be opened at one point to incorporate an extra processor which can boot the network. An easy way to implement the latter approach is to insert the extra processor into one row of the network. Two links from the new processor may be used to maintain the toroidal

```
VAL x.trans IS 4:                                   --x dimension size
VAL y.trans IS 4:                                   --y dimension size
... SC work(VAL INT i,j,x,y
         CHAN OF ANY left.in,left.out,right.in,right.out,
                    up.in,up.out,down.in,down.out)
[2][2][x.trans][y.trans]CHAN OF ANY link:
VAL horz IS 0:
VAL vert IS 1:
VAL a    IS 0:
VAL b    IS 1:
PLACED PAR
  PLACED PAR j = 0 FOR y.trans    --place processor in y dimension
    PLACED PAR i = 0 FOR x.trans  --place processor in x dimension
      VAL p IS i+(j*x.trans):               --processor number
      VAL m IS (i+1)\x.trans:     --next channel in x dimension
      VAL n IS (j+1)\y.trans:     --next channel in y dimension
      PROCESSOR p T8
        PLACE link[a][horz][i][j] AT 4:
        PLACE link[b][horz][i][j] AT 0:
        PLACE link[b][horz][m][j] AT 6:
        PLACE link[a][horz][m][j] AT 2:
        PLACE link[a][vert][i][j] AT 7:
        PLACE link[b][vert][i][j] AT 3:
        PLACE link[b][vert][i][n] AT 5:
        PLACE link[a][vert][i][n] AT 1:
        work (i,j,x.trans,y.trans,
              link[a][horz][i][j],link[b][horz][i][j],
              link[b][horz][m][j],link[a][horz][m][j],
              link[a][vert][i][j],link[b][vert][i][j],
              link[b][vert][i][n],link[a][vert][i][n])
```

Figure 2-6b
Configuration description for a toroidally connected network

structure and the rest are then available for booting. This is most easily accomplished by configuring the network with two PLACED PAR structures, one for a two-dimensional block of processors, and one for a ring of the toroid. The end processor of the ring must then be connected to the external processor.

Two-dimensional processor arrays like toroids are a very popular architecture for MIMD computers and have many advantages over simpler architectures such as rings, or more complex architectures such as trees or hypercubes. Two-dimensional arrays are infinitely scalable and map easily to a wide variety of computing problems and techniques. The number of links needed by each processor stays the same as the toroid grows. This fact makes scaling, connecting, and constructing a toroidal array quite easy.

As in the ring, however, interprocessor distance in a toroid becomes a constraint as the toroid grows. The maximum distance between any two processors in a toroid is the sum of one half the x and one half the y dimensions of the toroid. Thus, the amount of communication overhead varies with the square root of the number of processors in the toroid.

A ring is a subset of a toroid and can be created from a toroid in either of two ways: a one-dimensional slice of the toroid in either dimension can be taken, or the toroid can be considered as a "snake" with connected rows of processors moving alternately across and up the array. An even number of rows is needed to complete the ring and the last processor in the top row must use the wrap-around link to connect to the first processor on the bottom row. In the toroid shown in Fig. 2-6a, the embedded ring consists of processors 12-13-14-15-11-10-9-8-4-5-6-7-3-2-1-0.

A minor difficulty in working with toroids is that they cannot be scaled quite as flexibly as rings. To change the size of a toroid, we must add or subtract one full row or column at a time. Rows and columns can be added independently of each other, however, so the aspect ratio of a toroid can be changed at will. But, for a given number of processors, the less square the array, the greater the relative interprocessor distance will be. Therefore, creating toroids which are very much wider than high, or vice versa, should be done with caution.

Because toroids can be created in such a wide assortment of shapes, they can be used with many different styles of programming. Just as on the ring, both one- and two-dimensional pipeline structures are easy to implement on a toroid. And, in the same way that one-dimensional data structures map easily to rings, two-dimensional data structures distribute efficiently over toroids.

Hypercubes

Hypercube architectures are much more complex than either rings or toroids. A binary hypercube is a binary n-cube, with two processors on each edge of the n-cube, and where n is the cube's dimension and can represent any number greater than or equal to zero. If n is two, a square is created; if n is three, a cube is created; n equal to four defines a four-dimensional "cube," or hypercube, and so on. Each dimension larger than the previous requires twice as many nodes and one additional link for each node. Higher-order hypercubes which have more than two processors on each edge (for example, ternary hypercubes have three processors on each edge) require an extra link and cannot be constructed from the following example.

Figure 2-7a is an attempt to represent a four-dimensional cube on a two-dimensional page. This binary hypercube can be more easily imagined as two cubes, one nesting inside the other, with the corners of one cube connected to the corresponding corners of the second cube. Each corner is a node, and each node in a four-dimensional hypercube has four neighbors, the three found in a normal three-dimensional cube, plus the fourth node which connects to the other cube. The nodes of the outside cube are shaded darkly, while the nodes of the inside cube are shaded more lightly. The configuration description for this hypercube is listed in Fig. 2-7b.

The configuration description (Fig. 2-7b) begins on the first line with the separately compiled fold for the program which runs in each node of the hypercube.

Architectures
Chapter 2

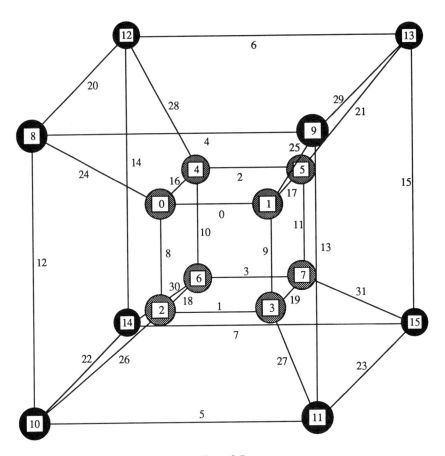

Figure 2-7a
A four-dimensional binary hypercube

Eight one-way links are passed to the program. These links are called d0.in, d0.out, and so on, continuing to d3.out. The d refers to dimensions zero through three. The next few lines of the description define some constants: dim is the dimension of the hypercube, n is two since we are creating a binary, multidimensional cube, size is one half the number of processors $n^{dim}/2$, and num.-links is the number of links. Since two processors must always share a point-to-point link, the number of links is always one-half the number of processors times the number of links per processor. The number of links per processor is always the dimension of the hypercube. Thus, the total number of links defined is $dim \times 2^{(dim-1)}$. A second dimension of channels must also be defined to provide for the input/output link exchange on each unidirectional link.

```
...   SC node(CHAN OF ANY d0.in,d0.out,d1.in,d1.out
                          d2.in,d2.out,d3.in,d3.out)
VAL n      IS 2:                                      -- binary hypercube
VAL dim    IS 4:                                      --number dimensions
VAL size IS (n<<(dim-1)):
VAL num.links IS size*dim:                            --32 links
[num.links][2]CHAN OF ANY a:
PLACED PAR k3 = 0 FOR n                                         -- dim 3
  PLACED PAR k2 = 0 FOR n                                       -- dim 2
    PLACED PAR k1 = 0 FOR n                                     -- dim 1
      PLACED PAR k0 = 0 FOR n                                   -- dim 0
        VAL n0 IS (((k3*(n*n))+(k2*n))+k1)+0:       --link dim 0
        VAL n1 IS (((k3*(n*n))+(k2*n))+k0)+size:    --link dim 1
        VAL n2 IS (((k3*(n*n))+(k1*n))+k0)+(size*2):--link dim 2
        VAL n3 IS (((k2*(n*n))+(k1*n))+k0)+(size*3):--link dim 3
        PROCESSOR ((((k3*(n*(n*n)))+(k2*(n*n)))+(k1*n))+k0) T8
          PLACE a[n0][k0]       AT 4:                 --left-right
          PLACE a[n0][1-k0]     AT 0:
          PLACE a[n1][k1]       AT 5:                 --up-down
          PLACE a[n1][1-k1]     AT 1:
          PLACE a[n2][k2]       AT 6:                 --front-back
          PLACE a[n2][1-k2]     AT 2:
          PLACE a[n3][k3]       AT 7:                 --fourth dim
          PLACE a[n3][1-k3]     AT 3:
          node(a[n0][k0],a[n0][1-k0],
               a[n1][k1],a[n1][1-k1],
               a[n2][k2],a[n2][1-k2],
               a[n3][k3],a[n3][1-k3])
```

Figure 2-7b
Configuration description of a binary hypercube

The binary hypercube architecture is actually defined in a somewhat recursive manner within the PLACED PAR structure. The configuration description (Fig. 2-7b) shows dim levels of nested PLACED PAR structures with the processor definition at the innermost level. Each level is replicated n times and defines the processors in one dimension. Since the configuration language does not directly support recursive structures, these four levels must be explicitly included and cannot be made in a PAR structure replicated dim times. Associated with each dimension is a constant equal to the correct channel number connecting the corresponding processor to its counterpart in that dimension. Making networks larger cannot be done simply by changing the value of dim; also required are an additional level of nesting, another value for connecting processors in the new dimension, and additional PLACE parameters.

To number the processors in the binary hypercube, we assign the processor's position in each dimension of the network to a bit in a four-bit, base-two number.

The processor numbers then range from 0 to 15 and correspond to the position, represented by a 0 or 1, of each processor in each of the four dimensions.

Once the processors in the hypercube are defined, the most difficult part of the configuration, ensuring that the links are connected properly, begins. The channel array a is defined as one large array (with a second dimension for exchanging the input/output pair) so that connecting the processors correctly becomes a matter of numbering the channels appropriately. Referring to Fig. 2-7a, consider the dimension from left to right across the page to be dimension 0, from top to bottom as dimension 1, in and out of the page to be dimension 2, and dimension 3, the "fourth dimension," to be the direction which connects the inner cube to the outer. Variables $k0$ through $k3$ in the PLACED PAR structures correspond to the dimensions 0 through 3.

The channels in this binary hypercube structure, elements of channel array a, are numbered in dimensional order. Those links which go from left to right (dimension 0) are numbered first, then those from top to bottom (dimension 1), and so on. To accomplish this numbering, the values $n0$ to $n3$ are used. Each PLACED PAR structure uses one of the values $k0$ through $k3$ to define n processors. These values indicate the position of that processor within the hypercube. The links are then numbered much as the processors are. While each of the dimensions corresponds to a bit in a four-bit, base-two number, the bit to which the dimension corresponds depends on which dimension the channels being numbered are in. Value $n0$ counts the dimension zero channels and does not depend on $k0$. Value $n1$ counts the channels in dimension one and adds to this count the number of channels already assigned. Values $n2$ and $n3$ do the same for dimensions two and three. The actual channels are then PLACEd with appropriate attention given to exchanging the input and output links. The dimension number of a link is the same as that link's hardware address. Figure 2-7a shows the processor number and channel element assignments for the hypercube.

The configuration file for the binary hypercube creates a closed network using all of the links of a four-link transputer, just as the toroid did. If the network must be booted from a link, the network must be opened. This, however, is much more difficult with the hypercube than it is with the toroid. To open a link on a four-dimensional hypercube, we must break down the hypercube into two three-dimensional cubes which are separately defined. These two cubes are then connected piecemeal, and an additional processor is inserted on an edge between processors of the two different three-dimensional cubes.

The binary hypercube architecture is very popular; several commercial parallel computers use it. It can be scaled to an arbitrary number of dimensions, but not without some difficulty. The network size in a hypercube can only be increased by doubling its size and adding an additional link to every processor. This requirement makes it difficult to make a practically scalable machine in a wide range of sizes. Either the nodes themselves must be replaced when the network size changes or some hardware will be wasted for the unused links in smaller networks.

Although they present greater hardware difficulties, binary hypercubes do offer better interprocessor connectivity than toroids. The greatest distance between any two processors in a binary hypercube varies as the logarithm of the number of processors. If a network doubles in size, the greatest interprocessor distance increases only by one. The total communication overhead in the computer, therefore, remains relatively small. The higher connectivity enjoyed by a binary hypercube gives it great flexibility.

A binary hypercube of dimension greater than four contains a toroid as a subset, just as a toroid contains a ring. Indeed, a four-dimensional binary hypercube is actually identical to a four-by-four toroid. The binary hypercube configured in Fig. 2-7b is thus a four-by-four toroid as well, although the link address assignments are different than in the toroid.

Hypercube architectures are most useful for distributed data applications. Since rings are a simple subset of a hypercube, it is very easy to pipeline data, but pipelining will generally waste much of the rich interconnectivity available in the hypercube. Because of the higher connectivity found in hypercube architectures, problems requiring much interprocessor communication can be done more efficiently on a hypercube than on a simpler architecture such as a toroid. Notable among these problems are frequency transforms. A Fourier transform, for example, can be partitioned on a binary hypercube so that data need never move farther than one processor.

Ternary Trees

Tree networks are very different from both toroids and hypercubes. Most architectures are homogeneous, but trees are layered structures which grow larger at lower levels. Trees typically begin with a single root processor at the top layer. This root processor is connected to a second layer of processors, each of which is a "child." Each child has its own "children" to which it is connected in the third layer, and the structure continues growing in this fashion. At the bottom layer of a tree, the nodes have no children. These childless nodes are sometimes called "leaves," and all of the higher nodes except the root node are called "branches," so that the entire tree structure resembles an upside-down natural tree. Figure 2-8a shows a ternary tree, a tree whose members each have three children, so that each layer has three times as many members as the layer above. In a binary tree, each processor has two children in the layer below it.

The configuration description for a ternary tree is given in Fig. 2-8b. This configuration actually describes a tree with one more layer (of twenty-seven nodes) than is shown in Fig. 2-8a. It includes three node programs, one for the root processor, one for the branch nodes, and the last for the leaf nodes. In the example, a node number and a set of channels are passed to each processor. The root node has no parent and as a result has six channels, two each for the three bidirectional links connected to its children. The branch nodes also include a pair of channels for the parent link, making eight channels in all. Only a processor number and the channels connected to its parent are passed to a leaf node.

Architectures Chapter 2

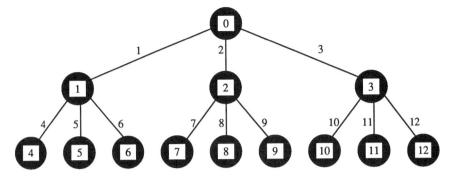

Figure 2-8a
Ternary tree network

```
...  SC root(VAL INT p,
         CHAN OF ANY child1.in,child1.out,
                     child2.in,child2.out,
                     child3.in,child3.out)
...  SC branch(VAL INT p,
         CHAN OF ANY parent.in,parent.out,
                     child1.in,child1.out,
                     child2.in,child2.out,
                     child3.in,child3.out)
...  SC leaf(VAL INT p,
         CHAN OF ANY parent.in,parent.out)
VAL width IS 3:
VAL node.count IS 40:
[node.count]CHAN OF ANY a,b:
VAL p1 IS 0:
VAL k1 IS 0:
VAL p2 IS ((p1+k1)*width)+1:
PLACED PAR                                        --level 0
  VAL p IS p1+k1:
  VAL n IS k1+1:
  PROCESSOR p T8
    PLACE a[n]     AT 5:                          --child 0 link
    PLACE b[n]     AT 1:
    PLACE a[n+1]   AT 6:                          --child 1 link
    PLACE b[n+1]   AT 2:
    PLACE a[n+2]   AT 7:                          --child 2 link
    PLACE b[n+2]   AT 3:
    root(p,a[n],b[n],a[n+1],b[n+1],a[n+2],b[n+2])
```

Figure 2-8b
Configuration description for a ternary tree network

```
    PLACED PAR k2 = 0 FOR width
      VAL p3 IS ((p2+k2)*width)+1:
      PLACED PAR                                      --level 1
        VAL p IS p2 + k2:
        VAL n IS p3:
        PROCESSOR p T8
          PLACE b[p]     AT 4:                        --parent link
          PLACE a[p]     AT 0:
          PLACE a[n]     AT 5:                        --child 0 link
          PLACE b[n]     AT 1:
          PLACE a[n+1]   AT 6:                        --child 1 link
          PLACE b[n+1]   AT 2:
          PLACE a[n+2]   AT 7:                        --child 2 link
          PLACE b[n+2]   AT 3:
          branch(p,b[p],a[p],
                 a[n],b[n],a[n+1],b[n+1],a[n+2],b[n+2])

        PLACED PAR k3 = 0 FOR width
          VAL p4 IS ((p3+k3)*width)+1:
          PLACED PAR                                  --level 2
            VAL p IS p3+k3:
            VAL n IS p4:
            PROCESSOR p T8
              PLACE b[p]     AT 4:                    --parent link
              PLACE a[p]     AT 0:
              PLACE a[n]     AT 5:                    --child 0 link
              PLACE b[n]     AT 1:
              PLACE a[n+1]   AT 6:                    --child 1 link
              PLACE b[n+1]   AT 2:
              PLACE a[n+2]   AT 7:                    --child 2 link
              PLACE b[n+2]   AT 3:
              branch(p,b[p],a[p],a[n],b[n],
                     a[n+1],b[n+1],a[n+2],b[n+2])

            PLACED PAR k4 = 0 FOR width               --level 3
              VAL p IS p4+k4:
              PROCESSOR p T8
                PLACE b[p]     AT 4:                  --parent link
                PLACE a[p]     AT 0:
                leaf(p,b[p],a[p])
```

Figure 2-8b (cont.)
Configuration description for a ternary tree network

The tree configuration includes two parameters, one that defines the number of descendants of each node (width), and a second which defines the total number of nodes in the tree (node.count). The node.count is the sum of as many powers of width as there are layers in the tree, counting the root node as layer zero. This calculation is easy to make for a binary tree; the number of nodes will be one less than two raised to the power of the number of layers. For a ternary tree with four layers, the number of nodes will be 1 + 3 + 9 + 27, equaling 40. The general expression for the number of nodes in a tree of k layers, each of whose nodes have n children is:

$$Number\ of\ nodes = \frac{n^k - 1}{n - 1}$$

The number of two-way channels needed to interconnect the nodes is the same as the number of nodes. Each node except the root node must have a channel connecting it to its parent. For this tree configuration, a parent channel is numbered with the same value as its processor.

In the configuration description for the ternary tree network, both a and b channel arrays are defined; each pair is used to do the input/output exchange of channels between pairs of processors. The parent channel's input must always be connected to the output of its children, and vice versa.

The tree configuration listed in Fig. 2-8b is organized as a series of nested layers, much like the hypercube configuration. Each configuration layer defines a layer of nodes with its children. The replicated nodes in each layer have two parameters associated with them, a p variable (p1 in layer one, p2 in layer two, etc.), and a k variable (k1 for layer one, k2 for layer 2, etc.). The p variable for each node defines the processor number of the first processor in the replicated set of three nodes, and k is used as the replicating variable in the PLACED PAR structure. Together, the p and k variables are used to construct p itself, the processor number of the node being defined and its associated parent channel element. Notice that the p variables for a layer are defined in the previous layer and are needed to connect each child's parent link (the node's child link). Parent links are put at address 0, and the children's links assigned to addresses in the order in which they are defined.

Each layer of PLACED PAR structures in the ternary tree network configuration has two members. The first of these implements the node itself, and the second implements the node's children using a replicated structure. This replicated PAR then constitutes the next layer. Each definition of a p variable uses the p value from the previous layer to keep track of how many nodes have been defined to this point. Processors in the bottom layer need not define a p value for their children, since they have none.

The configuration description for a tree can be enlarged to any extent as long as the nested structure is extended properly. In creating a larger tree, the value of node.count must be set properly and additional nested layers added. Since the number of links for each node stays the same regardless of the network size, a real network can use the same processor modules for a system of any size.

The interprocessor distance in trees generally varies with the logarithm of the number of nodes, just as the interprocessor distance for the hypercube does. The tree grows exponentially as layers are added, but the greatest possible distance from any leaf to another is twice the height of the tree.

As is true of hypercubes, tree networks are most useful for distributed data applications. It is difficult to create a sensible processor pipeline from a tree. Tree architectures are often used for applications in rule-based artificial intelligence and for searching large quantities of information. Trees are especially useful for answering questions made up of a hierarchy of subquestions which can be mapped onto a tree structure. Tree architectures can also be used to execute programs which begin with a large amount of raw information at the leaves and progressively reduce the amount of data while increasing the information content of the data at successively higher branches.

Trees do not often contain other regular networks as subsets, but other networks can contain trees. A tree embedded within a regular network, however, usually either is irregular or does not use every processor and link in the regular network. As an example of an embedded tree, one constructed by an exploratory program on a toroidal network (discussed in Chap. 7) is shown in Fig. 2-9.

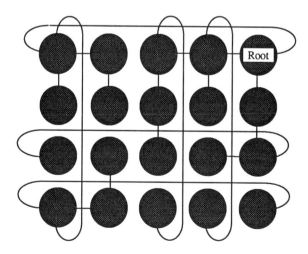

Figure 2-9
A processor tree as a subset of a toroidal network

In Summary

Transputers can easily be configured in a wide variety of useful and interesting architectures. These architectures can be elegantly described using the PLACED PAR structure and channel interconnections available in occam. In this chapter we have examined ring, toroid, hypercube, and ternary tree networks. Ring networks are the simplest and can be infinitely scaled by one processor at a time. However, the maximum interprocessor distance in a ring is significant, resulting in reduced performance efficiency. The more complex toroidal architecture can also be infinitely scaled, but only by one row or column at a time. The maximum interprocessor distance in a toroid is much less than in a ring. A hypercube network is yet more complex and more densely connected. It, too, can be scaled infinitely, but can only change in size by a factor of two. A ternary tree is a very different architecture, useful for data base searching and artificial intelligence applications.

Chapter 3

Processor Farms

The processor farm approach to parallel processing is one of the most flexible. This approach is generally independent of system architecture, can be used for almost any kind of problem with a parallel solution, and, once the basic software is set up, is very easy to use and adapt for a variety of purposes. In a sense, farming is considered to be parallel processing only in that farms require a controlling processor to direct other processors (worker nodes) and to organize communications between itself and the worker processors. The worker nodes themselves operate alone just as a traditional processor does.

A processor farm is a group of independent processors assigned tasks by a controller. Each processor performs its assigned task, returns the results of the task, and waits for new work (Fig. 3-1). Generally, the controller keeps track of which processors are busy or idle, what work has been completed, and what work remains.

To direct a processor farm, the controller must be able to communicate with each worker. Beyond this obvious requirement, there are virtually no restrictions on the actual hardware. Thus, practically any MIMD parallel processor can be used to implement a processor farm. Because communication between the processors is the only system requirement (other than the presence of the processors themselves), the transputer is an excellent candidate to use for processor farm systems. The transputer links provide a simple and efficient means of interprocessor communication, and the cpu itself is an effective distributed memory processor node.

Saying that any architecture or processor network can implement a processor farm does not imply that every network will perform equally well on any task. Processor farms generally receive all of their input data and instructions and output all of their results through their communication channels. This communication load can put a heavy burden on the channels. Thus, although almost any task *can* be performed using a processor farm, it may be more practical and efficient in some cases to use other approaches to distributing tasks and data. Generally, tasks that require a lot of work but little communication of data, and which do not generate large amounts of data to be returned to the controller, are good candidates for processor farms. Likewise, networks in which both the total amount of interprocessor communication hardware and the software required to support communication are relatively small will be most effective for use with processor farms. Transputer networks programmed in occam effectively meet both of these criteria.

Processor farms generally utilize identical processor nodes which do identical work. For the sake of simplicity, all of the following examples assume this to

Processor Farms Chapter 3

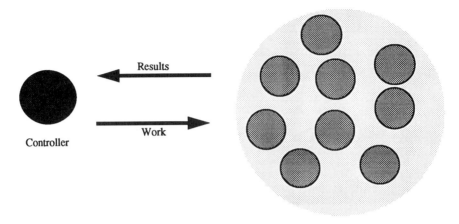

Figure 3-1
A processor farm consisting of a controller which communicates work
to a group of processors and receives results from the processors

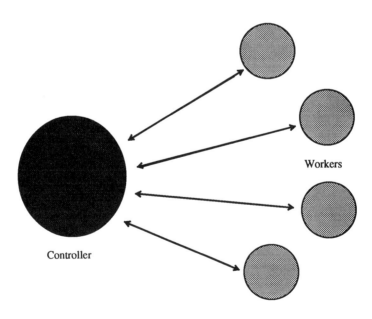

Figure 3-2
A processor farm with four processors connected directly to the controller

be the case. If this is not true for a given system, it will be difficult to port the software to larger or smaller systems of the same type.

A Simple Processor Farm

The simplest processor farm structure that can be envisioned using transputers is shown in Fig. 3-2. The farm controller can be directly connected to as many processors as it has links. Two programs must be written to organize a simple processor farm: one for the controller and one which is run by each of the workers in the processor array.

The workers have a very simple program (Fig. 3-3), one which reads instructions and data (called `work`), processes the data according to instructions in a routine called `do`, places the computed information into `results`, and then writes `results` back to the controller. At this point the worker awaits a new task. The channels `in` and `out` are local names for the channels which are connected to the controller. These names must be PLACEd to match the actual channel used in the hardware.

The program run by the controller is more complex than the worker program. Assuming that there are more tasks than workers, the simple control routine allocates work to the available processors, and stores the results (Fig. 3-4). The number of processors in the processor farm and the number of tasks to be performed are defined as shown. The tasks are stored in `tasks` and their results in `results`. The next variables defined, `tasks.done` and `tasks.assigned`, keep track of which tasks have been done and which remain to be assigned. The `in` and `out` channel arrays pass tasks to and receive work from the workers. These channels must be assigned to the correct hardware address with a PLACE statement (not shown).

The controller begins work by initializing the tasks, an application-specific job, and the variables `tasks.done` and `tasks.assigned` are cleared. Since none of the processors has any work at the beginning, it is safe to begin with the immediate assignment of `number.of.processors` tasks. These tasks are

```
CHAN OF INT in, out:
INT work, results:
...Process do(results, work)
...channel placement
WHILE TRUE
  SEQ
    in? work
    do(results, work)
    out! results
```

Figure 3-3
Code for a worker in a simple processor farm

```
VAL number.of.tasks IS 100:
VAL number.of.processors IS 4:
[Number.of.tasks]INT results, tasks:
INT tasks.done,tasks.assigned:
[number.of.processors]CHAN OF INT in,out:
SEQ
  ...Initialize task array
  tasks.done:=0
  tasks.assigned:=0
  PAR i=0 FOR number.of.processors        --start up workers
    out[i]!tasks[i]
  tasks.assigned:=number.of.processors
  WHILE tasks.assigned < number.of.tasks  --while tasks remain
    ALT i=0 FOR number.of.processors      --listen for any worker
      in[i]?results[tasks.done]           --input results
        SEQ
          out[i]!tasks[tasks.assigned]    --send new work
          tasks.assigned:=tasks.assigned+1
          tasks.done:=tasks.done+1
  PAR i=0 FOR number.of.processors        --empty the workers
    in[i]?results[tasks.done+i]
```

Figure 3-4
Code for a processor farm controller

passed out, and as the workers begin computing, `tasks.assigned` is incremented. The controller then enters an interactive phase, waiting for work to be returned on any channel. This is accomplished with an ALT which waits for any input, reads the input into the `results` array, assigns new work to the worker, and updates the `tasks.assigned` and `tasks.done` pointers. Note that the work performed may not be completed and stored in the same order in which the tasks were assigned.

When `tasks.assigned` equals `number.of.tasks` there is no more work to be assigned, but there is still data being produced by each of the workers. The controller then waits for the remaining data from each worker. The code in Fig. 3-4 shows this being done with a PAR structure and the data being assigned to the `results` buffer.

Efficiency Concerns

This simple example illustrates two basic efficiency concerns about processor farm programs: the assignment and receipt of tasks by the controller, and the efficient distribution of tasks through the network.

The controller assigns tasks and receives results from the processor farm within an ALT; it could also, however, be received within a PAR, potentially making use of four links at once rather than just one. This is a much more efficient use of links, but it can cause some problems with work distribution and load balancing.

For maximum efficiency in a processor farm, every processor should do the same amount of work; that is, the work load should be evenly balanced among the processors. In the first example, the tasks originated from a single process in the controller which assigned the tasks on a first-come, first-served basis. This procedure is reasonably efficient since, if one job is more difficult than another, the processor doing the easier job can receive another task before the processor with the more difficult task is finished. Thus there is inherent load balancing in the approach, unless there are only a few tasks per processor so that the whole job doesn't have time to balance out.

If PARs are used in the controller, the task assignment can be done somewhat differently. One approach is to distribute the work among the PARs at the beginning; each PAR would then essentially be an independent program doing a completely independent task (Fig. 3-5). This amounts to apportioning the work ahead of time and does not allow for load balancing of the work. If one of the processors falls behind, there is no way to transfer some of its work to another processor.

Figure 3-6 shows another solution to the problem of task assignment. If another process is constructed within the controller, all of the links can be employed at once to communicate with the workers while the work assignment is still load-balanced. In this solution, however, the communication bottleneck has simply been moved back one stage. The process feeding the four PARs will now handle requests for data one at a time. This may be better than the original approach (a single-process controller), however, since the feeder process has less work to do and the internal communications are very fast. A strict comparison of approaches would require an actual implementation with work and result buffers of a well-defined size.

Another load-balancing issue affecting the efficient distribution of tasks must be dealt with when multiprocessor farms are used. Earlier we alluded to problems involved in having only a few tasks per processor. If one task is substantially more difficult than the others, the processor with that task will continue to work long after the other processors have finished. This is clearly inefficient. Since the goal is to have every processor working for the same amount of time, there must be enough tasks so that if the tasks are assigned on a first-come, first-served basis to each of the processors, all of the processors will finish at the same instant.

It is not easy to ensure this happy state. If every task is identically difficult, the processing load will be perfectly balanced if the number of tasks is a multiple of the number of processors. Every processor will do the same number of identical tasks and finish at the same time. Since it is not often the case that the tasks are the same and that their number divides evenly by the number of processors, we must look further.

If the tasks become more and more varied in their difficulty (although the average difficulty remains the same), more and more tasks will be needed to average

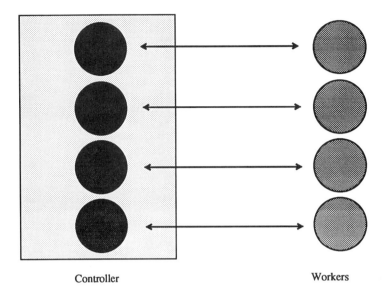

Figure 3-5
A processor farm controller running with four parallel processes, one for each worker

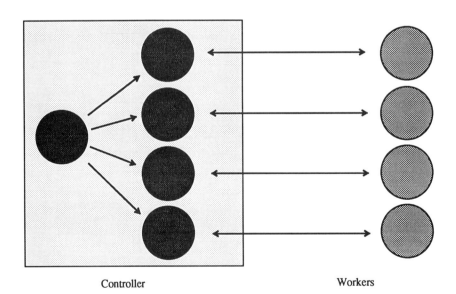

Figure 3-6
A processor farm controller running with five parallel processes, one for each worker, and one to distribute the tasks among them

out the random fluctuations in task difficulty and maintain load balancing. Thus as the task difficulty becomes more varied, the ratio of number of tasks to number of processors must increase for an even work distribution over the processors to be maintained.

Solving any specific problem on a parallel machine requires that the problem be divided up into tasks, each of which can be done by a processor. If the size of each task is very small, the problem is said to be fine-grained; if each task is large, the problem is said to be large-grained. The granularity of a problem, then, is a measure of the size of the tasks into which a problem is divided. Computers are also often referred to as large- or fine-grained depending upon whether they have a few very fast and powerful processors (the typical supercomputer) or many small, less powerful processors (an SIMD machine, for example). Computers between the two extremes are called medium-grained.

Clearly, if a problem is inherently large-grained, and if the difficulty of each subtask is very different from that of other subtasks, it will be difficult to balance the computing load using a processor farm. Many problems, however, can be divided up into increasingly smaller pieces and the granularity of the tasks can, within limits, be arbitrarily decided. Such problems can be handled well with processor farms.

It is easy to assume from this discussion that the finer the grain of a problem, the more efficiently it can be processed; but this is not necessarily true. It is certainly true that fine-grained tasks are easier to load balance, but there is more to efficiency on a parallel computer than load balancing. As we remarked in the first chapter, one of the fundamental limits of parallel processing is the communication overhead. As the number of tasks increases, the communication required to set up the task, pass it to a processor, and return the results increases as well. Thus there is an optimum granularity for any given problem. The optimum granularity is the granularity which minimizes the communication overhead and maximizes the load balancing. Figure 3-7 shows an idealized curve representing the relationship between granularity and efficiency.

Generally, the maximum efficiency for a particular computer is achieved when the granularity of the problem matches the granularity of the computer. Large-grained computers need large-grained tasks and small-grained computers need small-grained tasks. The granularity of a computer is determined by the ratio of its computing power to its communications capability. Since the transputer is a single-chip microprocessor with significant communications capability, it qualifies as a medium-grained computer. Thus a processor farm composed of transputers should generally be given a collection of medium-grained tasks.

Large Processor Farms

Processor farms are typically made of many more than just the five processors shown in Fig. 3-2. In any larger system, all of the nodes must still communicate with the controller to receive data and return results. If there are insufficient

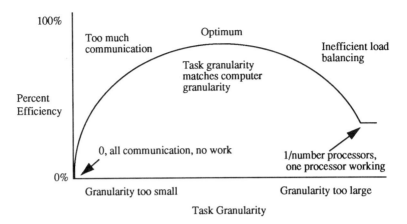

Figure 3-7
Idealized plot of task granularity versus efficiency for a given parallel processor

links to make a direct connection between each processor node and the controller (as is always the case for larger networks), there are two alternatives open to the designer.

In the first alternative, the network can be switched so that every processor is connected directly to the controller (although not all at the same time). A second alternative is for the processors to communicate through other processors which are connected to the controller. Figure 3-8 shows such a tree structure of processors connected to a controller through interprocessor links. Processors lower in the tree must receive or pass messages through those higher in the tree. This tree can be made as large as is desired, but the taller the tree is the farther messages must pass between controller and processors. A wider, shorter tree has a smaller communication distance between processor and controller, but the width is constrained by the number of communication links available to each processor.

A switched network can use the programs described in Figs. 3-3 and 3-4 with some additions for switch controlling. The actual hardware, of course, will be much more complex than with a simple tree of link interconnections. On the other hand, a network with direct, unswitched links is more complex because it uses an interprocessor message-passing scheme to communicate with the controller.

In an unswitched processor tree like the one shown in Fig. 3-8, each node is connected to one processor above it, usually termed the *parent*. The nodes connected below are called *children*. Thus each processor has one parent and zero or more children (Fig. 3-9).

Since each processor node is running asynchronously, every processor with children must be prepared to pass data up or down the tree at any moment. Figure 3-10 shows a logical schematic of such a processor node with two children. The

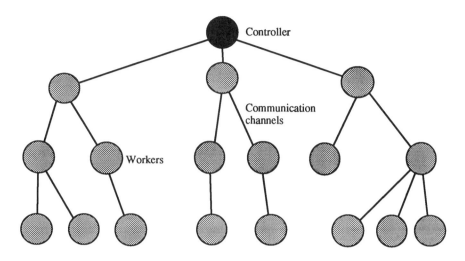

Figure 3-8
A processor tree with a controller at the top and worker nodes
connected by channels in a tree below

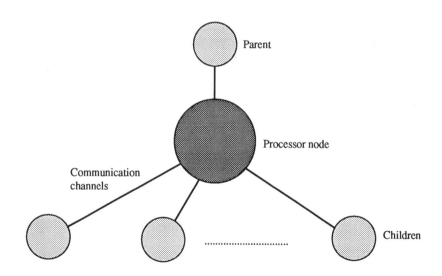

Figure 3-9
A processor node in a simple tree with one parent and zero or more children
connected by communication channels

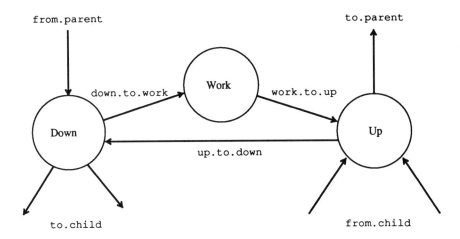

Figure 3-10
A logical block diagram of a processor node with three processes: one for communications down the tree, one for communications up the tree, and one to do work

input/output processes must run totally independently and at a higher priority than the local worker process in order to guarantee that communications up and down the tree are never delayed. If the input/output processes are not at a higher priority and independent of the worker process, the worker process will slow down the communications. This in turn will slow down the assignment of tasks and the delivery of results, leaving some processors farther down the tree with nothing to do.

In Fig. 3-10, there are three processes shown running in parallel: one process performs work locally (the "work" process), a second process inputs information from a parent and passes it either to the local worker or to a child (the "down" process), and the third process reads information from the children or the worker and passes it to the parent (the "up" process). Figure 3-11 shows the program organization. Of the three processes, the worker process is the simplest (Fig. 3-12). This process waits for work to appear on its local channel down.to.work, processes it, passes the results through the local channel work.to.up, and then awaits more work.

The down process (Fig. 3-13) is the most complex. It distributes data received from the parent to the appropriate channel. To accomplish this distribution, the down process must keep a record of which processors are free, and it must have some method for deciding to which processor the work must go if more than one processor is available. For this example, a very simple algorithm is used for making the decision. The down process will try to pass the work to the subtree with the greatest number of free processors in link order; if the subtrees are full, the work is passed to the local work process. Work will thus tend to be passed to the bottom of the tree. If there are fewer tasks than processors, using this approach will result

```
CHAN OF INT up.to.down, down.to.work, work.to.up,
CHAN OF INT from.child1, from.child2, to.child1, to.child2:
CHAN OF INT from.parent, to.parent,
PRI PAR
  PAR
    ...down process                              Fig. 3-13
    ...up process                                Fig. 3-14
  ...worker process                              Fig. 3-12
```

Figure 3-11
Code for worker node with down, up, and worker routines

```
WHILE TRUE
  INT work, results:
  SEQ
    down.to.work? work
    ...do(work,results)
    work.to.up!    results
```

Figure 3-12
Code for work process in a worker node in which work is read in, processed, and sent out

```
INT count1,count2,work,link:
SEQ
  up.to.down? count1;count2
  WHILE TRUE
    ALT
      from.parent? work
        IF
          count1 > count2
            SEQ
              to.child1! work
              count1:=count1-1
          count2 > 0
            SEQ
              to.child2! work
              count2:=count2-1
          TRUE
            down.to.work!work

      up.to.down? link
        IF
          link=1
            count1:=count1+1
          link=2
            count2:=count2+1
```

Figure 3-13
Code which passes work down the tree

```
INT count1,count2,results:
SEQ
  from.child1?  count1
  from.child2?  count2
  to.parent!    (count1+count2)+1
  up.to.down!   count1;count2
  WHILE TRUE
    SEQ
      ALT
        work.to.up?     results
          SKIP
        from.child1?    results
          up.to.down! 1
        from.child2?    results
          up.to.down! 2
      to.parent!      results
```

Figure 3-14
Code which passes data up the tree

in unnecessary communication, but if there are many tasks, this algorithm will have the tendency to leave processors high on the tree without work and free to devote themselves to communication. Since processors higher on the tree must do more communication than those lower down (because there are more results produced below them to pass up), this algorithm should have the effect of increasing overall efficiency.

A simple comparison shows which subtree has the greatest number of free processors. If there is no subtree with a free processor, work will be sent to the local work process. If the various subtrees are unbalanced, this algorithm will have the effect of filling the largest tree first. When the number of free processors in the largest subtree equals the number of processors in the next largest subtree, the work will be shared between the two subtrees, and so on, until all of the subtrees are receiving equal shares of work.

The down process (Fig. 3-13) keeps a count of free workers in the subtree connected to each channel. Once the count of these workers is initialized (more on that later), the down process simply subtracts one from the available processor count for the appropriate channel whenever a job is passed to that channel. If that count reaches zero, a different channel must be used. With this arrangement, the controller must never pass work if there is not a free worker available to perform it.

When results are passed up the tree, the up process becomes involved. The up process very simply waits for an input from any child processor or the local worker, passes the input results to the parent, and reports to the down process that another worker is available (Fig. 3-14). The down process must then increment the count of available processors for that channel.

In order for each processor to obtain the initial count of available processors beneath it on the tree, the tree must calculate its own size. A simple way to perform this calculation is to have each processor pass a message up the tree to the controller. The message is a count of processors in the subtree to which the channel is connected. The up process in each node inputs a count from each of its children, passes the count to the down process which records it, sums and increments the count from all the children, and passes the total to the parent processor. This procedure repeats in every processor, and a total count bubbles up to the controller. Obviously a processor at the bottom of the tree, with no children, will simply output a one, representing itself.

Storage and Communication Issues

One of the drawbacks to a farming program like the one just described is the amount of storage space required. Because each of the three processes is run in parallel, each must maintain a separate memory space. Both the down and worker processes must allocate storage for the task data, and both the worker and up processes must allocate storage for the results data. For small data sets this storage space requirement will not matter much, but if the data sets become very large, it might be necessary to sacrifice some communications capability or processor asynchronicity to keep the programs running.

A second drawback to this farming scheme is the communication overhead. A significant amount of time can be wasted while work and results pass up and down the tree. When a processor completes a task it must wait until the results travel all of the way up the tree to the controller, competing all the while for the use of the link with other results. It must then wait for a task to make its way back down. For a processor low on the tree, this wait may be lengthy. Clearly, the link traffic can form a bottleneck.

To minimize this problem, individual processors can queue work locally. Then while the results of the first task are making their way to the processor, work can begin on the queued task. To implement this queue, we can insert a high-priority buffer between the down and worker processes. When the tree first initializes the count of processors, each processor must count itself as (size of queue + 1) workers, rather than one worker as before. There are, however, two problems with this approach. First, the storage requirements within each processor must be increased, and second, when there is no more work to be distributed to a network, it is possible that some processors may have queued tasks remaining while other processors have no work at all.

Notice that processors high on the tree will spend proportionately more time doing input and output than processors lower on the tree. If the results are not needed in any particular order this work distribution is not an issue, and the communication can be run at a higher priority than the work, so that the network maintains a reasonable efficiency.

A Real-World Example

Consider a practical example illustrating this discussion, the calculation of the Mandelbrot set over a 512 by 512 array of points using a processor farm. This example is a popular exercise and easily adaptable to farming methods on a parallel computer.

The Mandelbrot set (named for the mathematician Benoit Mandelbrot) is the set of points in the complex plane which are quasi-stable when iterated in a function. The most commonly used function is $z = z^2 + c$, with the complex value z initially equal to zero and the constant c the point being tested in the complex plane. After the first iteration, z will be equal to c. The stability of a point in the plane can be measured by iterating the function until its magnitude exceeds an arbitrary limit, usually two. The number of iterations (up to some arbitrary maximum) required to exceed the limit is the stability. If the function does not exceed this limit, the point in the plane is considered to be in the Mandelbrot set. This calculation is popular because it is a simple test of raw processor power and because the image produced by assigning colors to the stability levels for a two-dimensional set of points is exceedingly strange and beautiful.

```
PROC man.calc.64(INT count, VAL REAL64 p.real,p.imag)
  VAL INT max.count IS 255:
  VAL REAL64 max.size IS 4.0 (REAL64):
  REAL64 z.real,z.imag,z.size,t.real,t.imag:
  SEQ
    count   := 0
    z.real  := p.real
    z.imag  := p.imag
    t.real  := z.real*z.real
    t.imag  := z.imag*z.imag
    z.size  := t.real+t.imag
    z.size:=z.size-max.size
    INT64 test RETYPES z.size:
    WHILE (count<max.count) AND (test < 0 (INT64))
      SEQ
        z.imag :=((z.real+z.real)*z.imag)+p.imag
        z.real := (t.real-t.imag)+p.real
        t.real := z.real*z.real
        t.imag := z.imag*z.imag
        z.size := t.real+t.imag
        z.size:=z.size-max.size
        count   :=count+1
:
```

Figure 3-15
Mandelbrot calculation routine

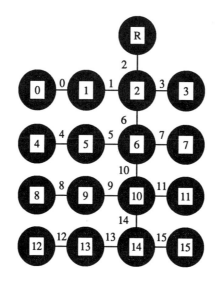

Figure 3-16
A mesh used as an unbalanced tree

The procedure used to actually calculate the stability of a point is shown in Fig. 3-15. This procedure takes a complex value (p.real, p.imag) and returns an iteration count; the computation is done with 64-bit floating point arithmetic. The maximum number of iterations is set at 255, and the magnitude threshold for z squared is set at four. The z and t values are used as local temporary variables.

The stability calculation routine first initializes count, assigns the complex p values to z, and calculates the magnitude squared minus the threshold (stored in z.size). This magnitude must be compared to the max.size limit. So that a floating-point comparison is avoided, z.size is first retyped to a 64-bit integer and then compared to zero to test the threshold limit. The routine then enters a WHILE loop predicated on a check of the iteration count and the threshold limit. If neither limit is exceeded, the loop simply calculates the function and magnitude until either the count or the threshold is exceeded. The library mandel_lib.tsr includes this entire routine.

When calculating the Mandelbrot set with a processor farm, we must distribute the computation routine to every processor. Each processor in the farm must calculate a two-dimensional block of points, receiving the complex values for the block from its parent and returning an array of bytes which are the iteration values for the function at each point in the block.

The network for this processor farm (Fig. 3-16) is a simple, unbalanced tree connected to a root node and built of a subset of the connections available in a mesh-connected array of processors. The configuration of this network is shown in Figs. 3-17 and 3-18. Figure 3-17 simply defines two channel protocols, one for passing data down the tree, and the second for passing results up the tree. The i.size and r.size values determine the size of the block of points to be calculated. Generally, in this program, variables prefixed with *i* represent imaginary values and variables prefixed with *r* represent real values. The code in Fig. 3-17 is stored in a library called protocol_lib.tsr.

Figure 3-18 shows the configuration description of the network. It includes the protocol library and the first line of each procedure for the various processors in the network. The root node (shown with an R in Fig. 3-16) is first, followed by the processors with zero, one, two, or three children. Only the link connections

```
VAL INT i.size IS 32:
VAL INT r.size IS 32:
PROTOCOL DOWN IS INT;INT;REAL64;REAL64;REAL64;REAL64:
PROTOCOL UP
  CASE
    count; INT
    data;  INT;INT;[r.size][i.size]BYTE
:
```

Figure 3-17
Protocol and block size definition for the Mandelbrot set calculation

```
#USE "protocol_lib.tsr"
PROC node.controller(CHAN OF UP     from.network,
                     CHAN OF DOWN   to.network)
PROC no.child(    CHAN OF DOWN   from.parent,
                  CHAN OF UP     to.parent)
PROC one.child(   CHAN OF DOWN   from.parent,
                  CHAN OF UP     to.parent,
                  CHAN OF UP     from.child1,
                  CHAN OF DOWN   to.child1)
PROC two.child(   CHAN OF DOWN   from.parent,
                  CHAN OF UP     to.parent,
                  CHAN OF UP     from.child1,
                  CHAN OF DOWN   to.child1,
                  CHAN OF UP     from.child2,
                  CHAN OF DOWN   to.child2)
PROC three.child(CHAN OF DOWN    from.parent,
                  CHAN OF UP     to.parent,
                  CHAN OF UP     from.child1,
                  CHAN OF DOWN   to.child1,
                  CHAN OF UP     from.child2,
                  CHAN OF DOWN   to.child2,
                  CHAN OF UP     from.child3,
                  CHAN OF DOWN   to.child3)
VAL x.trans IS 10:
VAL y.trans IS 12:
VAL num.trans IS x.trans*y.trans:
VAL num.left  IS x.trans/2:
VAL num.right IS x.trans-(num.left+1):
[num.trans]CHAN OF DOWN d.link:
[num.trans]CHAN OF UP   u.link:
PLACED PAR
  PROCESSOR 999 T4                                        --root node
    PLACE u.link[num.left]   AT 7:
    PLACE d.link[num.left]   AT 3:
    node.controller(u.link[num.left],d.link[num.left])
```

Figure 3-18
Configuration file for the tree network

```
PLACED PAR k=0 FOR (y.trans-1)             --replicate top rows
  VAL i IS (k*x.trans):
  PLACED PAR
    VAL p IS i+num.left:
    PROCESSOR p T8                         --center node for top rows
      PLACE d.link[p]            AT 5:
      PLACE u.link[p]            AT 1:
      PLACE u.link[p+x.trans]    AT 7:
      PLACE d.link[p+x.trans]    AT 3:
      PLACE u.link[p-1]          AT 4:
      PLACE d.link[p-1]          AT 0:
      PLACE u.link[p+1]          AT 6:
      PLACE d.link[p+1]          AT 2:
      three.child(d.link[p],         u.link[p],
                  u.link[p-1],       d.link[p-1],
                  u.link[p+1],       d.link[p+1],
                  u.link[p+x.trans], d.link[p+x.trans])
    PLACED PAR
      VAL p IS i:
      PROCESSOR p T8                       --end node for left side
        PLACE d.link[p]    AT 6:
        PLACE u.link[p]    AT 2:
        no.child(d.link[p],u.link[p])
      PLACED PAR p=(i+1) FOR (num.left-1)       --left side
        PROCESSOR p T8
          PLACE u.link[p-1]  AT 4:
          PLACE d.link[p-1]  AT 0:
          PLACE d.link[p]    AT 6:
          PLACE u.link[p]    AT 2:
          one.child(d.link[p],   u.link[p],
                    u.link[p-1], d.link[p-1])
      VAL q IS i+num.left:
      PLACED PAR
        VAL p IS q + num.right:
        PROCESSOR p T8                     --end node for right side
          PLACE d.link[p]    AT 4:
          PLACE u.link[p]    AT 0:
          no.child(d.link[p],u.link[p])
        PLACED PAR p=(q+1) FOR (num.right-1)     --right side
          PROCESSOR p T8
            PLACE u.link[p+1]  AT 6:
            PLACE d.link[p+1]  AT 2:
            PLACE d.link[p]    AT 4:
            PLACE u.link[p]    AT 0:
            one.child(d.link[p],   u.link[p],
                      u.link[p+1], d.link[p+1])
```

Figure 3-18 (cont.)
Configuration file for the tree network

```
VAL i IS ((y.trans-1)*x.trans):
PLACED PAR                                       --bottom row
  VAL p IS i+num.left:
  PROCESSOR p T8                  --center node for bottom row
    PLACE d.link[p]          AT 5:
    PLACE u.link[p]          AT 1:
    PLACE u.link[p-1]        AT 4:
    PLACE d.link[p-1]        AT 0:
    PLACE u.link[p+1]        AT 6:
    PLACE d.link[p+1]        AT 2:
    two.child(d.link[p],       u.link[p],
              u.link[p-1],     d.link[p-1],
              u.link[p+1],     d.link[p+1])
PLACED PAR
  VAL p IS i:
  PROCESSOR p T8                      --end node for left side
    PLACE d.link[p]     AT 6:
    PLACE u.link[p]     AT 2:
    no.child(d.link[p],u.link[p])
  PLACED PAR p=(i+1) FOR (num.left-1)          --left side
    PROCESSOR p T8
      PLACE u.link[p-1]    AT 4:
      PLACE d.link[p-1]    AT 0:
      PLACE d.link[p]      AT 6:
      PLACE u.link[p]      AT 2:
      one.child(d.link[p],     u.link[p],
                u.link[p-1],   d.link[p-1])
VAL q IS i+num.left:
PLACED PAR
  VAL p IS q + num.right:
  PROCESSOR p T8                     --end node for right side
    PLACE d.link[p]    AT 4:
    PLACE u.link[p]    AT 0:
    no.child(d.link[p],u.link[p])
  PLACED PAR p=(q+1) FOR (num.right-1)          --right side
    PROCESSOR p T8
      PLACE u.link[p+1]    AT 6:
      PLACE d.link[p+1]    AT 2:
      PLACE d.link[p]      AT 4:
      PLACE u.link[p]      AT 0:
      one.child(d.link[p],     u.link[p],
                u.link[p+1],   d.link[p+1])
```

Figure 3-18 (cont.)
Configuration file for the tree network

needed to construct the tree of processors are passed as arguments to each routine. Channels passing data down the tree use the DOWN protocol, and those returning results use the UP protocol. The x and y dimensions of the array of processors are defined with x.trans and y.trans. By changing these values, we can make the network dimensions and aspect ratio larger or smaller.

Each row of the tree is constructed in three parts: a left side, a right side, and a center processor. The left and right sides are constructed with a replicated PAR and have a processor with no children on their ends. The other processors each have one child. The center processor connects the two sides and has three children. Each row is then replicated to create a deeper tree; the bottom layer is identical with those above it except that its center processor has two children. The configuration description defines the root node first, followed by a replicated structure of rows. In each row, the center processor is defined first, then a replicated structure for the left side of the row, and last, a replicated structure for the right side of the row. Finally, the bottom row is defined (center first, then left side, then right side) with a two-child central processor.

In a tree structure, there are as many bidirectional channels needed to connect the processors as there are processors. Therefore, the channels are created with an array, each of whose elements are associated with the processor of the same number. Figure 3-16 shows the processor and channel numbering. The physical processor links are actually connected with link 0 on the left, link 2 on the right, link 1 above, and link 3 below. Notice that the parent channel is associated with link 2 on the left side of each row but link 0 on the right side. Two arrays of channels are actually defined to accommodate the input and output for each of the links. Channel names prefixed with *u* are used for links which transmit results up the tree, and channel names prefixed with *d* pass data down.

Two of the procedures which run in the processor nodes are shown in Figs. 3-19 (zero children) and 3-20 (three children). The procedures for one and two children are simply stripped down versions of the three-child routine. All of these procedures use the same methods described earlier, but with one minor difference: the initial count of processors in the tree is done before the PAR structure is entered.

The child routine begins with library definitions. Next, the channels are defined with the UP and DOWN protocols which describe the information passed on the channels. The information passed down is a work assignment and comprises 6 values, an x and y pair which are not needed for the computation but which identify the resulting data, and four real values. X and y give the address (in the 512 by 512 array) of the upper left value of the block of values being generated. The first two 64-bit real values identify the first point to be calculated in the complex plane; the second two values represent the increment to be added in the real and imaginary directions when the next point is calculated.

The UP protocol has two cases. In the first case an INT is used to define the communication protocol for passing the count up the tree at the beginning of the program. In the second case the protocol used to pass calculated data back up to

```
#USE "mandel_lib.tsr"
#USE "protocol_lib.tsr"
PROC no.child(CHAN OF DOWN from.parent,
              CHAN OF UP to.parent)
  CHAN OF DOWN down.to.work:
  CHAN OF UP   work.to.up:
  SEQ
    to.parent! count; 1
    PRI PAR
      PAR
        REAL64 real.start,imag.start,r.delta,i.delta:
        INT x,y:
        WHILE TRUE
          SEQ
            from.parent ?  x;y;real.start;imag.start;
                           r.delta;i.delta
            down.to.work!  x;y;real.start;imag.start;
                           r.delta;i.delta
        [r.size][i.size]BYTE image:
        INT x,y:
        WHILE TRUE
          SEQ
            work.to.up? CASE data;   x;y;image
            to.parent!          data;   x;y;image

      [r.size][i.size]BYTE image:
      REAL64 r.current,i.current,r.delta,i.delta,r.start:
      INT count,x,y:
      WHILE TRUE
        SEQ
          down.to.work? x;y;r.start;i.current;
                        r.delta;i.delta
          SEQ i=0 FOR i.size
            SEQ
              r.current:=r.start
              SEQ r=0 FOR r.size
                SEQ
                  man.calc.64(count,r.current,i.current)
                  r.current:=r.current+r.delta
                  image[i][r]:=BYTE count
              i.current:=i.current+i.delta

          work.to.up! data; x;y;image
:
```

Figure 3-19
Code for processor with no children

```
#USE "mandel_lib.tsr"
#USE "protocol_lib.tsr"
PROC three.child(CHAN OF DOWN from.parent,
                 CHAN OF UP   to.parent,
                 CHAN OF UP   from.child1,
                 CHAN OF DOWN to.child1,
                 CHAN OF UP   from.child2,
                 CHAN OF DOWN to.child2,
                 CHAN OF UP   from.child3,
                 CHAN OF DOWN to.child3)
  INT  num.free.1,num.free.2,num.free.3:
  CHAN OF INT   up.to.down:
  CHAN OF DOWN  down.to.work:
  CHAN OF UP    work.to.up:
  SEQ
    from.child1? CASE count; num.free.1
    from.child2? CASE count; num.free.2
    from.child3? CASE count; num.free.3
    to.parent!        count; (num.free.1+num.free.2)+
                                     (num.free.3+1)

    PRI PAR
      PAR
        REAL64 real.start,imag.start,r.delta,i.delta:
        INT x,y:
        WHILE TRUE
          ALT
            from.parent ?  x;y;real.start;imag.start;
                              r.delta;i.delta
              IF
                (num.free.1>num.free.2)AND(num.free.1>num.free.3)
                  SEQ
                    to.child1!  x;y;real.start;imag.start;
                                   r.delta;i.delta
                    num.free.1:=num.free.1-1
                num.free.2>num.free.3
                  SEQ
                    to.child2!  x;y;real.start;imag.start;
                                   r.delta;i.delta
                    num.free.2:=num.free.2-1
                num.free.3>0
                  SEQ
                    to.child3!  x;y;real.start;imag.start;
                                   r.delta;i.delta
                    num.free.3:=num.free.3-1
                TRUE
                  down.to.work! x;y;real.start;imag.start;
                                   r.delta;i.delta
```

Figure 3-20
Code for processor with three children

```
        INT link:
        up.to.down? link
          IF
            link=1
              num.free.1:=num.free.1+1
            link=2
              num.free.2:=num.free.2+1
            TRUE
              num.free.3:=num.free.3+1
  [r.size][i.size]BYTE image:
  INT x,y:
  WHILE TRUE
    SEQ
      PRI ALT
        work.to.up?  CASE data; x;y;image
          SKIP
        from.child1? CASE data; x;y;image
          up.to.down! 1
        from.child2? CASE data; x;y;image
          up.to.down! 2
        from.child3? CASE data; x;y;image
          up.to.down! 3
      to.parent!         data; x;y;image
  [r.size][i.size]BYTE image:
  REAL64 r.current,i.current,r.delta,i.delta,r.start:
  INT count,x,y:
  WHILE TRUE
    SEQ
      down.to.work? x;y;r.start;i.current;
                         r.delta;i.delta
      SEQ i=0 FOR i.size
        SEQ
          r.current:=r.start
          SEQ r=0 FOR r.size
            SEQ
              man.calc.64(count,r.current,i.current)
              r.current:=r.current+r.delta
              image[i][r]:=BYTE count
          i.current:=i.current+i.delta
      work.to.up! data; x;y;image
:
```

Figure 3-20 (cont.)
Code for processor with three children

the root node is defined. It comprises both the x and y values mentioned earlier, and a two-dimensional byte array of results. The internal channels down.to.-work and work.to.up are also defined with the DOWN and UP protocols respectively. The up.to.down channel simply passes an integer value and uses an INT protocol.

The code for the processor nodes with no children is much simpler than the code for those with children. The no.child routine (Fig. 3-19) does not need the ALT input in the up code, the up.to.down channel, or the calculation and testing of the number of free children below it on the tree. Initializing this no.child routine is straightforward and simply consists of a one being passed to the processor's parent. The three-child routine uses the same structure described earlier and is shown in Fig. 3-20.

Other than the differences in the channel protocols and the actual work done in the work routine, this real-world example is substantially similar to the example in the earlier discussion of farms. As with the earlier example, the work code itself is identical in every processor and simply reads in the task, calculates results, and passes the results on. The actual work of the Mandelbrot calculation begins with an initialization of the current real and imaginary values defining a point in the complex plane, and continues with a calculation of the stability. A doubly nested loop in the real and imaginary dimensions serves to iterate the current values appropriately, using the delta parameters passed from the controller. The loop count is the same as that of the block of points being calculated, i.size by r.size. The count returned from the man.calc.64 call is converted to a byte and stored in the image array. When all of the values have been computed, the data together with the x and y identifiers are output on channel work.to.up, and the processor begins waiting for another task.

The code running in the network controller for the Mandelbrot example is diagrammed in Fig. 3-21. Two parallel processes basically constitute the controller: one for receiving instructions from the host and sending tasks to the network (send-work), and a second process for reading results from the network and passing the data back to the host (get-results). These two processes are connected by the internal channel ready. To begin the program, the host passes four real values to the send-work process. The first pair of values defines the first and last boundary points to be examined in the real dimension, and the second pair of values defines the first and last values in the imaginary dimension. The number of points to be computed is defined as a value in the routine itself. From this information, process send-work can define each of the tasks to be performed. These tasks are passed to the network whenever the second process sends a "ready" command over channel ready.

The get-results process in the root node controls the input of results and the output of "ready" commands. When the network is initialized, a count of the number of processors available in the network is passed up the processor tree and read in by get-results. Tasks can be immediately assigned to each of the processors in

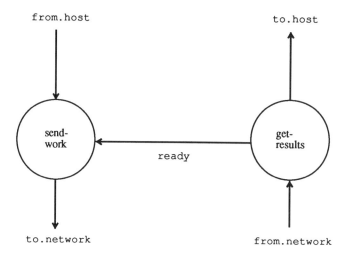

Figure 3-21
A logical block diagram of the controlling root node

the farm, so get-results passes `tree.size` signals to send-work, which passes tasks on to the network. Thereafter, every time a task is finished and some tasks remain to be done, get-results sends another signal to send-work. The data itself is passed on to the host. By decoupling the assignment of work and the receipt of data in this way, the network controller avoids bottlenecks in the flow of information either to the host or to the network.

The network control code itself is listed in Fig. 3-22. The controller is connected to the network through the `to.network` and `from.network` channels, which are defined with the UP and DOWN protocols just like the channels in the network itself. The host channels do not use a protocol and are PLACEd at link 0. Channel `ready` is an internal channel which passes only dummy integer values. After the channel definitions, the image size and a calculation of the number of blocks (tasks) are listed. Of course, the number of blocks depends directly on the block size defined in the protocol library.

After the variables and channels are defined, the network control program begins with an infinite WHILE TRUE loop by reading the starting values from the host and calculating the incremental values for the address of the first point in each block (`r.delta` and `i.delta`). It then calculates the incremental value for each point in the real and imaginary planes themselves (`i.inc` and `r.inc`). The send-work process then initializes the current point values (`r.current` and `i.current`), enters a double loop which moves the current values through the correct sequence of points in the complex plane, and waits for a "ready" signal from process get-results. The incremental values passed to the network remain the same for every task, whereas the address of the first point in each block must be updated (by

```
#USE "protocol_lib.tsr"
PROC node.controller(CHAN OF UP     from.network,
                     CHAN OF DOWN to.network)
  CHAN OF ANY to.host,from.host:
  PLACE from.host AT 4:
  PLACE to.host   AT 0:
  CHAN OF INT ready:
  VAL INT x.image   IS 512:
  VAL INT y.image   IS 512:
  VAL INT r.blocks IS x.image/r.size:
  VAL INT i.blocks IS y.image/i.size:
  PAR
    INT x.start,x.end,y.start,y.end,r.block,i.block:
    REAL64 r.current,i.current,r.start,i.start,r.end,i.end:
    REAL64 r.inc,i.inc,r.delta,i.delta:
    WHILE TRUE
      SEQ
        from.host? r.start;r.end;i.start;i.end
        r.delta  := (r.end-r.start)/(REAL64 ROUND r.blocks)
        i.delta  := (i.end-i.start)/(REAL64 ROUND i.blocks)
        r.inc    := r.delta/(REAL64 ROUND r.size)
        i.inc    := i.delta/(REAL64 ROUND i.size)
        i.current:=i.start
        SEQ i=0 FOR i.blocks
          SEQ
            r.current:=r.start
            SEQ r=0 FOR r.blocks
              INT link:
              SEQ
                ready? link
                to.network! r;i;r.current;i.current;
                            r.inc;i.inc
                r.current:=r.current+r.delta
            i.current:=i.current+i.delta

    VAL INT num.blocks  IS r.blocks*i.blocks:
    VAL INT image.size   IS r.size*i.size:
    [r.size][i.size]BYTE image:
    INT tree.size:
    SEQ
      from.network? CASE count; tree.size
      IF
        tree.size > num.blocks
          tree.size:=num.blocks
        TRUE
          SKIP
```

Figure 3-22
Code for network controller in the Mandelbrot example

```
        WHILE TRUE
          SEQ
            SEQ i=0 FOR tree.size
              ready! 0
            INT x,y:
            SEQ i=0 FOR num.blocks
              SEQ
                from.network? CASE data; x;y;image
                PRI PAR
                  IF
                    i < (num.blocks-tree.size)
                      ready! 0
                    TRUE
                      SKIP
                  to.host! x;y;image.size;image
            to.host! 0;0;0
:
```

Figure 3-22 (cont.)
Code for network controller in the Mandelbrot example

r.delta and i.delta) with each new block assigned. After r.blocks times i.blocks tasks have been assigned, the entire job is finished and send-work awaits a new assignment from the host.

The get-results process begins by reading in the network size (tree.size). If the number of tasks is less than the network size, however, the size variable is set equal to the number of tasks. This change prevents process get-results from initiating more tasks than are available. The process then enters an infinite loop and sends tree.size signals to process send-work, which starts up the tasks. Data may now be available from the network, and if available will be read in by get-results. Each time a task is completed and read in, process get-results must pass a signal to process send-work to initialize another task, unless there are no tasks remaining (number of tasks received is less than the network size). At the same time, the results read from the network are returned to the host. The signal to send-work and the output to the host are done in a PRI PAR structure so that a delay on one channel will not delay the other. The signal on channel ready has priority, on the assumption that it is more important to start new work than to deliver old results. When all of the tasks have been completed, get-results passes three zeros to the host to indicate the end of the work.

Efficiency Measurements

The calculation of the Mandelbrot set for a 512 square set of values was actually run on a variety of network sizes and the performance of the processor farm measured for a variety of block sizes. The block size is a measure of the granularity of the task. If larger blocks are used, fewer are needed, and the granularity of the

Processor Farms Chapter 3

Block size	Computation time for tree size (seconds)				Number of blocks
	10 x 12	7 x 9	5 x 6	3 x 4	
1 x 1	32.77	32.71	32.66	45.31	262144
2 x 2	8.60	9.50	14.79	34.17	65536
4 x 4	3.50	6.36	13.05	32.24	16384
8 x 8	3.33*	6.18*	12.79*	31.78	4096
16 x 16	3.58	6.43	12.95	31.75*	1024
32 x 32	4.75	7.33	13.88	32.39	256
64 x 64	10.71	10.71	18.21	34.64	64
128 x 128	42.17	42.15	42.14	45.41	16
256 x 256	143.20	143.13	143.09	143.05	4
512 x 512	379.82	379.52	379.36	379.21	1

*Optimum granularity

Figure 3-23
Timing measurements for various tree sizes and granularity

	Efficiency of optimum block size for tree size				
	10 x 12	7 x 9	5 x 6	3 x 4	
Efficiency	.950	.975	.989	.994	

Figure 3-24
Computational efficiency of various networks at optimum granularity

problem increases; smaller blocks result in a smaller granularity. The network dimensions were chosen to maintain a consistent aspect ratio over a large range of network sizes. Figure 3-23 is a table of the timing measurements. The center four columns show the times in seconds for the array size printed at the top of the column. The granularity of the problem increases towards the bottom of each column.

The program described constructs a 512-by-512 array of values by computing many smaller blocks which together make up the full 512 square array. A 1-by-1 block size (listed at the top of the left column) requires 262144 blocks (listed at the top of the right column) to complete the array, while a 16-by-16 block size requires 1024 blocks. Both the block size and the number of blocks define the granularity. The granularity ranges from the smallest possible (a block with one element) to the largest possible (one block constituting the entire calculation). Although all of the blocks are the same size for a given computation, some blocks require more effort to calculate than others.

An examination of the table entries reveals some interesting points. Very large granularity problems are inefficient on a processor farm since many of the processors are unused. Very small granularity problems are likewise inefficient since each calculation requires so much communication overhead. The optimal efficiency is found somewhere in the middle with a block size of 8-by-8 elements per task, a size somewhat smaller than one might expect. This observation points out the importance of keeping processors busy even if higher communication overhead results. The optimum block size is about the same for every tree size, although a close inspection of the measurements reveals that the optimum granularity for larger networks is slightly smaller than the optimum granularity for smaller networks. For example, the 12-processor network has an optimum granularity with blocks of size 16-by-16 while the 120-processor network has an optimum granularity with block sizes of about 8-by-8. The efficiency of each network at optimal granularity is shown in Fig. 3-24.

It is also interesting to note that passing data to the bottom of the network tree has only a very small effect on the performance of the system, even when the granularity of the problem is very large and there may be many free processors high up in the tree. For any given granularity, those networks with more processors than tasks perform nearly identically, regardless of the size of the network. For example, if the computing task is divided into four blocks of 256-by-256 elements each, the performance of the different networks varies by only about one tenth of a percent.

Figure 3-25 is a graph of the efficiency for each of the networks with various task granularity. The efficiency is the time for one processor to do the calculation (379 seconds) divided by the product of the number of processors in the network and the time for the network to do the calculation. This graph nicely corresponds with the idealized curve shown in Fig. 3-7 and demonstrates that smaller networks tend to be less sensitive to the granularity of the problem. This is obviously true if one considers a single processor network which gives the same efficiency for any

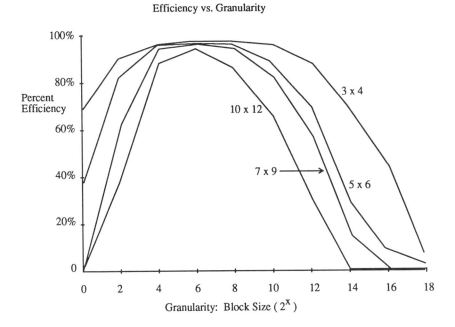

Figure 3-25
Plot of efficiency versus granularity for four network sizes and ten block sizes

granularity and would be plotted as a flat line at 100% efficiency. Also as expected, the smaller networks are more efficient at any granularity; there simply is not as much communication overhead.

In Summary

A processor farm is a simple parallel processing approach appropriate to nearly any MIMD computer. Under the guidance of a controlling processor, each processor in the network receives instructions, performs work, and returns results. If the network is switched, each processor can communicate directly with the controller, and very simple programs can be used to implement the farm. If an unswitched network is used, communication must take place through intervening processors in the network. In this case, special communication processes must be included in each processor to pass instructions and data to and from the controller. For an unswitched network, the drawbacks of communication overhead and redundant storage requirements are exchanged for the advantage of simplified hardware.

In an optimized processor farm, the granularity of each task will be matched to the granularity of the computer itself, and the overall computing task will be distributed evenly over the network. If the tasks are too large to be evenly distributed, some processors may not have enough work. If the tasks are too small, communication overhead will come to dominate the computer's performance. If a parallel program has only a few tasks which vary greatly in difficulty, the work may be unevenly distributed. Either the tasks must be broken up into smaller pieces which can be spread more evenly over the processor network or they must be made more consistent in difficulty.

Because the communication overhead in a processor farm can be troublesome, processor farms are most useful for tasks which require a relatively large amount of computing compared to communication. While any parallel machine is most efficient when running programs with a high ratio of work to communication, it is especially important that programmers using processor farms be careful of the amount of communication required for each task accomplished.

Chapter 4

Pipeline Processing

A parallel pipeline system is one in which every data element is passed through every processor with each processor performing a different operation on each of the data elements. This approach is sometimes called algorithmic parallelism because the algorithm tasks are distributed among multiple processors. In contrast, data parallelism (Chap. 5) is a method in which the entire program is placed in every processor and data is distributed among the processors.

Parallel pipeline systems are among the simplest of parallel systems to create. They are simple to understand, construct, control, and program. The communication techniques needed for pipelining are easy to develop and the programs easy to write. Because of their regular structure and simple data flow, pipeline systems are especially useful for real-time systems. But, although pipeline systems are easy to create and use, they do have some serious drawbacks. Because it is often difficult to distribute a processing task over a pipeline, pipeline systems can be problematic to use efficiently and to apply to many kinds of problems. Therefore, although useful in many situations, pipeline parallelism is suitable for a limited variety of applications.

Perhaps the most common pipeline systems are linear arrays and rings. A linear array of processors is shown in Fig. 4-1; the data flow and the interprocessor links are illustrated with arrows. In Fig. 4-2, the data flow and task distribution are shown. A linear structure is used for most of the examples of pipeline processing presented in this chapter, and will often be referred to as a *pipe*.

This chapter continues with a discussion of program issues and pipeline efficiency. Three communication methods with different buffering schemes for pipes are presented together with a demonstration illustrating the advantages and disadvantages of each. The chapter concludes with a discussion and example of multi-dimensional pipeline systems.

Program Issues

Pipeline methods tend to be fairly inflexible because the data must pass sequentially through a set of processors with specified operations performed on the data in each processor within the pipeline. Since the program is distributed over the processors in the pipe, the programmer must find some way to divide the work into different tasks, each of which can be performed at a different stage in the pipe. Because the data moves from one processor to the next, no processor can proceed until the previous processor in the pipe has completed its task and passed the data

Figure 4-1
A linear pipeline of processors

to it. This means that if one processor has less work than another, the faster processor will simply have to wait for the slower. In order to guarantee efficient use of processors in a pipeline, a programmer must ensure that every processor in the pipeline has exactly the same amount of work. If the tasks performed within the different processors of the pipeline are very different, accomplishing this can be quite difficult.

Because it is difficult to distribute a program over a pipeline efficiently, the program and pipeline size cannot be easily changed. Any change in the pipeline size requires a reallocation of tasks over the pipe to accommodate the difference in pipeline size. Unless the overall program can be divided into many small tasks which can easily be moved from one processor to another, this reallocation is likely to be difficult. Such inflexibility in program and pipeline size can be a considerable drawback in parallel systems for which scalability is important.

If each data element can be processed independently, there is one simple programming technique which can be implemented on a pipeline and which does maintain pipeline scalability. This technique is similar to distributed data parallelism but uses a pipeline communication structure. In this approach, data is passed sequentially through every processor, as is the case in any pipeline, but rather than performing one specific operation on every data element, each processor performs all of the processing on some of the data elements. All of the processors together will process every piece of data. Each processor except the first will then read in some processed and some unprocessed data. This approach is clearly scalable, since the amount of data processed in each processor can easily be changed to match the number of processors in the pipeline.

In addition to dealing with program allocation issues, a programmer using pipeline processors must be careful to minimize the amount of data passed between processors in order to lower the communication and storage overhead in the pipe. When processing data, many programs require extra storage for intermediate variables needed to compute the final result. It is important to organize the computation over the pipeline so that it is unnecessary to pass these intermediate variables

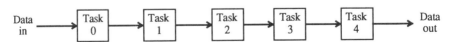

Figure 4-2
Task and data distribution for a pipeline of processors

Chapter 4   *Pipeline Processing*

between processors. This consideration adds another constraint to the distribution of a program over a pipeline.

Pipeline Efficiency

The efficiency of parallel pipeline systems depends first and foremost upon the distribution of the program over the multiple processors. The program must be divided into as many parallel pieces as there are processors, and each piece should require the same amount of work. The following analysis of pipeline efficiency assumes that this distribution has been done perfectly. Beyond program distribution, the efficiency of a pipe will depend on the number of processors in the pipe in relation to the amount of data to be processed, and the efficiency with which the processors can communicate.

One of the disadvantages of pipeline systems is that it is not possible for an entire pipe to begin working immediately on a set of data. Since the data is processed sequentially by each processor in turn, a particular processor in the pipe will not get any work until all of the previous processors in the pipe have finished working on the first data set. The same problem occurs at the end of the whole task as the first processors complete their work and the last processors are still hard at work. The overhead of filling and emptying the pipeline decreases the efficiency of the parallel system. Of course, if the work going on in a pipe continues for a very long time, the overhead of starting up and emptying the pipe becomes insignificant.

This overhead is easy to calculate for a theoretical parallel pipeline with no interprocessor communication overhead. The performance of such a theoretical system will be the maximum possible for a processor pipeline. A practical system, of course, will encounter some interprocessor communication overhead, and the system's performance will fall short of this theoretical limit.

Any set of data to be processed in a pipeline can be divided into packets, each of which is communicated as a group with one instruction. Consider the extreme case of a pipeline network processing all of its data in one packet. Each processor will work in turn, but since there is only one packet, only one processor can work at a time. Figure 4-3 is an activity diagram representing the activity in each processor for a five-processor pipe working with one packet. In this figure, a row of boxes represents the activity of each processor for each of the labelled time periods. A filled box represents a working processor for the time period, while an empty box represents an idle processor for the time period. The efficiency of the pipe shown is then 20%, since only one processor in five is working at a time (five time periods used out of 25 available). If n processors are used, the efficiency is $1/n$.

If this same data set is divided into three packets, the efficiency improves (Fig. 4-4). In this case, every processor works for three time periods; each time period is one third as long as the time periods in Fig. 4-3 because there is only one third as much data in each packet. The overall efficiency is now 43%, since the pipe is busy for 15 of the 35 time periods available. This pipeline will therefore process the data in half the time of the single-packet pipeline.

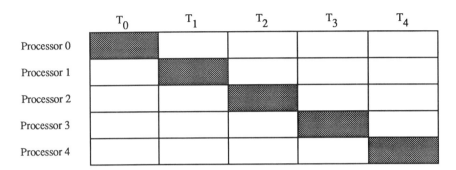

Figure 4-3
Activity diagram for a five-processor pipeline with one data packet

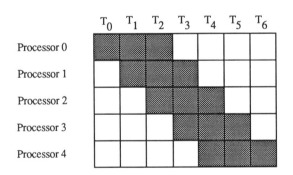

Figure 4-4
Activity diagram for a five-processor pipeline with three data packets

Parallel Programs for the Transputer

To completely fill or empty a pipe of n processors requires n time periods. Therefore, the total wasted time for starting and stopping a pipe of n processors is n − 1 periods per processor, or n × (n − 1) time periods for all of the processors in the pipe. The total wasted time in filling and emptying a pipe can be reduced by dividing the data into smaller packets, each of which can be processed more quickly. The smaller the packets, the more quickly the pipe is completely filled and the better the overall efficiency. Obviously, there is a lower limit of one element to the size of a data packet.

The number of time periods required to completely process a task will depend on the number of data packets into which the task is divided and the number of processors in the pipe. Again, if we ignore communication overhead, given N packets, P processors, and T seconds to process the slowest stage, the total processing time can be expressed as:

$$T_{total} = (N + (P - 1)) \times T_{slowest\ stage} \qquad \text{Equation 4-1}$$

The overall efficiency of a pipeline system is defined as the time to do the task with one processor, divided by the time for a pipeline processor to do the task times the number of processors. The processing time for the pipeline processor is the total time from Eq. 4-1. Given N packets, P processors, and T seconds to process the slowest stage, the overall efficiency can be expressed as:

$$E = \frac{T_{total\ with\ one\ processor}}{P \times (N + P - 1) \times T_{slowest\ stage}} \qquad \text{Equation 4-2}$$

If we assume that the problem is perfectly distributed over the pipeline (every stage has the same amount of work) and that there is no communication overhead needed to pass data through the pipeline, the equation can be simplified. The total time on one processor can then be expressed as the processing time per stage for one packet times the product of the number of packets and the number of stages (the number of processors). Given T seconds to process each packet at each stage, this can be expressed as:

$$E = \frac{N \times P \times T}{P \times (N + P - 1) \times T} \quad \text{or} \quad \frac{N}{N + P - 1} \qquad \text{Equation 4-3}$$

Equation 4-3 represents the theoretical limit of efficiency for a pipeline processor. In real life, communication overhead and task distribution problems will keep a practical system from reaching that limit so that the time in the numerator and the denominator cannot be considered as equal, and the equation cannot be simplified. That is, the time in the denominator must include the effects of interprocessor communication overhead.

This equation demonstrates two things. First, the more packets into which the data sets are divided, the greater the efficiency becomes. This makes sense, since the overhead of loading and unloading the pipe becomes less important as the number of packets becomes larger. Second, the efficiency is greater with fewer processors; this is also reasonable, since shorter pipelines with fewer processors do not take as long to load and unload and thus have lower overhead.

Theoretically, in order to maximize pipeline efficiency, the data packets should be made as small as possible. Practically, this is often the case, but it is also true that with smaller packets the communication overhead in the processors themselves becomes relatively greater, thereby reducing the overall efficiency. This trade-off between communication overhead and pipeline efficiency is hardware- and program-dependent, as is demonstrated in the next section.

Programmers should also note that if a pipeline is intended to run for a long time, the number of data packets will be very large and the overhead of starting up becomes insignificant. In that case, the packet size can be made as large as desired to reduce the communication overhead between processors.

A Pipeline Example

Consider, as an example, a 20-processor pipeline which must process 32,768 data elements. Each element in this example requires 1 millisecond of processing. With a perfect distribution of the problem over the pipe and with no communication overhead, the processing would take 32,768 divided by 20, or 1638.4 milliseconds. As we have seen, however, loading, unloading, and communicating data to the pipeline imposes overhead on the system. The program used to demonstrate this is the first example given later in the section on communication methods (Fig. 4-7).

Figure 4-5 presents the measured performance of the 20-processor pipeline. The number of packets in each test is shown in the left column and the consequent packet size in the right column. For each test, the communication overhead, the actual processing time (including any overhead), and the best performance theoretically possible are presented.

To measure the communication overhead for the pipeline, we run the test without any processing. This overhead includes the time needed to fill and empty the pipe, as well as the time required for communicating the data from one processor in the pipe to the next. The processing time is the actual measured computing time, including the time needed to fill the pipe, process all data, and communicate data from one stage in the pipe to the next. The processing task itself is generated artificially and can be perfectly distributed among the processors in the pipe. The best time is a calculation of the theoretically best performance that the pipeline could achieve given the number of packets, the length of the pipe, and the work to be done. To find this number, we divide the processing time on a perfectly distrib-

Number of packets	Computation time for each packet size (milliseconds) 32768 elements with 1 millisecond processing each			Packet size
	Communication time only	Total time	Best time	
1	1650	34419	32768	32768
2	933	18135	17203	16384
4	574	9996	9421	8192
8	395	5923	5530	4096
16	306	3890	3584	2048
32	261	2872	2611	1024
64	238	2364	2125	512
128	228	2110	1881	256
256	222	1984	1760	128
512	221	1923	1699	64
1024	221	1896	1669	32
2048	224	1890	1654	16
4096	231	1903	1646	8
8192	245	1938	1642	4
16384	273	2016	1640	2
32768	330	2164	1639	1

Figure 4-5

This table shows the performance measurements for various packet sizes on a ring with 20 processors. Each four-byte data element requires 1 millisecond of processing. The communication time shows only the time needed to pass the data through the pipe without any processing, and demonstrates the communication overhead; the total time is the actual measured performance; and the best time is the fastest the pipe could theoretically process the data if there were no communication overhead.

uted 20-processor pipe (1638.4 milliseconds) by the efficiency of the pipeline (found from Eq. 4-3).

A comparison of the measured performance and the theoretical limit of performance in Fig. 4-5 readily demonstrates that, in the real world, communication overhead is a significant factor in pipeline processing. The assumption made in simplifying Eq. 4-3 (no significant communication overhead) is inaccurate. A more accurate representation of the situation, explicitly including the effect of the communication overhead, is:

$$E = \frac{T_{total\ on\ one\ processor}}{T_{communication} + P \times (N + P - 1) \times T_{slowest\ stage}} \qquad \text{Equation 4-4}$$

The communication time for the 20-processor pipeline is shown in the second column of Fig. 4-5. When the communication time is subtracted from the total time in the third column, the result agrees very well with the best possible processing time shown in column four. The agreement is not quite as good for tasks with small packet sizes (smaller than 16 elements per packet). This fact can be explained if we note that the processor overhead of function calls, process switching, and so on, is relatively greater for small packet sizes so that the true processing time per element is substantially greater. This effect will be more pronounced for processing tasks which are small compared to the program overhead.

A closer inspection of the communication times shown in Fig. 4-5 also shows that the overhead depends upon the packet size. This is easily explained by noting that, with only a few large packets, the pipeline can only use a few links at a time, while with many small packets the overhead required to set up the packet communication becomes significant. This means that, for a given pipe length and number of data values to be processed, there is an optimal packet size. For this example, the most efficient packet size is about 32 words since the communication overhead measurement for 32-word packets is the smallest.

As we noted before, however, the more packets a pipeline processes, the better efficiency the pipeline can achieve. Thus the optimal processing time (including both the communication overhead and the pipeline overhead) is achieved with 2048 slightly smaller packets of 16 words each.

The efficiency of this pipeline example is plotted in Fig. 4-6 together with the theoretical limit of efficiency calculated from Eq. 4-3. This graph shows that the best efficiency actually achieved is less than 90%. Various simple programming methods that can improve the efficiency and decrease the communication overhead in a pipeline are discussed in the next section.

Communication Methods

Obviously, in any kind of multiprocessor system, including pipelines, each processor must communicate with its neighbor in order to pass data to it, and this

Chapter 4 Pipeline Processing

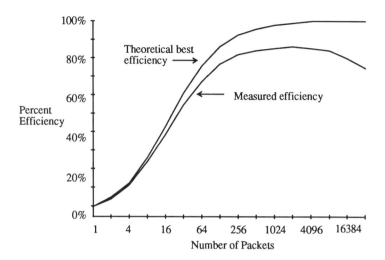

Figure 4-6
Number of data packets versus efficiency for a 20-processor pipe

```
VAL PacketSize IS 100:
PROC pipe(CHAN OF [PacketSize]INT left.in,right.out)
  ...    PROC work(data)
  [PacketSize]INT data:
  WHILE TRUE
    SEQ
      left.in?    data
      work(data)
      right.out!  data
:
```

Figure 4-7
A pipeline processor with a single buffer for input and output

Pipeline Processing   Chapter 4

communication represents overhead that a single processor does not encounter. It is important that this overhead be made as small as possible in order for a pipeline's efficiency to be as great as possible. The preceding efficiency analysis did not consider different methods of communication between the processors in a pipeline. The communication methods used will, of course, affect the pipeline's efficiency. Illustrated in the following examples are different methods for performing the communication between each processor and its neighbor as an infinite flow of data moves through the pipeline. The configuration description for all of the examples is similar to the ring described in Chap. 2.

Single Buffering

Single buffering, the simplest method for passing data in a pipeline, is programmed in Fig. 4-7. In this routine, each processor in the pipe has a single buffer, or memory space, for storing data. The processor first reads the data in, processes the data with process work, and then outputs the result. This very simple buffering structure has the drawback of performing the input and output sequentially with the work itself; it takes no advantage of the transputer's ability to do communication at the same time as processing. Thus the total processing time for any given processor is the sum of the time needed to do the input, the time to do the work, and the time to do the output.

A detailed activity diagram for this simple process using single buffering is shown in Fig. 4-8. Each row shows the activity of the labelled processor in each time period. In each box, I represents an input, P represents processing, O represents output, and an empty box represents no activity; the subscript value indicates which data packet is involved. Each processor in turn does an input, some processing, and an output. The output of one processor is done at the same time as the input for the following processor. The processing box is drawn twice the size of the input or output boxes to signify that the processing takes twice the time of the input or output. (The factor of two is chosen for consistency with the double-buffered ex-

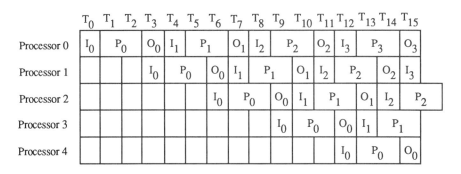

Figure 4-8
Detailed activity diagram for a five-processor, single-buffered pipeline

ample.) Thus the total processing rate for the pipe is one data packet in every four time periods. Only the first 16 time periods are shown.

It is clear from the diagram that the five-processor pipe shown in Fig. 4-8 will not output its first packet of data until time period 15. The 15-period delay is the latency of the pipeline, the time actually needed to process a single data packet. Although the processing rate for each stage is four periods per cycle and there are five processors in the pipe for a product of 20, the latency is only 15 periods since the input of one stage overlaps with the output of the previous stage, giving an effective delay of three periods per packet.

Double Buffering

A double-buffered pipeline (Fig. 4-9) is more efficient than the single-buffered pipeline. A double-buffered communication method uses two buffers so that data can be processed in one buffer while the second is used for input and output. This doubles the data storage requirement but allows communication and process-

```
VAL PacketSize IS 100:
PROC pipe(CHAN OF [PacketSize]INT left.in,right.out)
  ...  PROC work(data)
  [2][PacketSize]INT data:
  INT p0,p1,temp:
  SEQ
    p0:=0                                        --initialize pointers
    p1:=1
    [PacketSize]INT work.data IS data[p1]:
    [PacketSize]INT io.data   IS data[p0]:
    SEQ                                          --start pipe up
      left.in?   work.data                         --first input
      PAR
        left.in?  io.data                          --second input
        work(work.data)                            --first work
    WHILE TRUE
      SEQ                                        --main pipeline loop
        [PacketSize]INT work.data IS data[p0]:
        [PacketSize]INT io.data   IS data[p1]:
        PAR
          SEQ                                    --input and output
            right.out! io.data
            left.in?   io.data
          work(work.data)                                   --work
        temp:=p0                                 --exchange pointers
        p0:=p1
        p1:=temp
:
```

Figure 4-9
A pipeline processor with a double buffer for work and communications

Pipeline Processing   Chapter 4

ing to proceed at the same time. The total delay for one stage of the pipe is now the slowest of either the processing, or the input plus the output. In this example, the processing takes just the same time as the input and input together, so the total processing rate for a double-buffered pipe is two time periods for each data packet compared to four periods per data packet in the single-buffered pipe.

The double-buffered program uses a two-dimensional array to store the data and two pointers to alternately select each of the buffers. Initially, the pointers are arbitrarily set to point to one or the other of the buffers; at the end of each processing cycle these pointers are exchanged so that the pointer to the input and output buffer points to the work, and vice versa. Each processing cycle comprises, in parallel, a work process, and an output followed by the input. Initially, an input followed by a parallel input and work process is done to load the pipe for the first two processing cycles. The initial loading does not include an output, so it will also prevent an initial output of invalid data.

Figure 4-10 is an activity diagram for the double-buffered pipeline. The activity of each processor is shown in two rows of boxes; the top row represents the interprocessor communication and the bottom row represents the processing. N data sets must be output to fill a buffer in each processor of an n-length pipe before any data sets are output. If the initial input before the main loop were not included in the program, the first n data sets output would be invalid, but a data set would be received for every data set loaded, simplifying the control of the pipe.

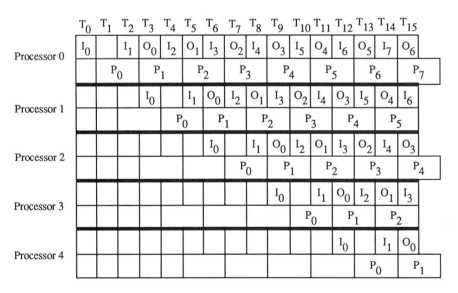

Figure 4-10
Detailed activity diagram for a double-buffered five-processor pipeline

100   *Parallel Programs for the Transputer*

Since a double-buffered pipe has two buffers at every stage, the latency is effectively doubled. Every data packet stays in each processor for two complete cycles or, in this example, four time periods. Again, since the input and output between two processors occur at the same time, the effective delay is one time period less, for a total delay of three time periods per processor. The five-stage pipe will thus have a total *latency* of 15 time periods (three periods per packet) while it processes data packets at the *rate* of one packet every two time periods.

Note that although the processing is shown in parallel with the communication in this example, the two processes could be organized differently. If the amount of data to be communicated grows larger or smaller as it passes to the next processor so that the input and output are not the same size, or if the work process becomes faster than either the input or output process, it can make sense to have the work process also use its buffer for the faster communication process. The point is that when using double buffering with input, output, and work processes in one node, the two fastest processes should share a buffer.

Triple Buffering

A double-buffered approach to pipelining is effective as long as some work needs to be done on the data being passed the pipe. However, if the processing time is less than the sum of the input and output time, triple buffering can be used to improve the pipeline efficiency even further. A triple-buffered pipeline is a pipeline with three buffers in each processor, one for input, one for output, and one for the work process.

Figure 4-11 shows the activity diagram for a triple-buffered pipe; each processor has three rows representing the input process at the top, the work process in the middle, and the output process at the bottom. The program finishes processing a data packet in each time period (except the first two) although each packet stays in each processor for three time periods. Again, the latency per processor is one less than three because of the overlap between input and output for communicating processors. The overall latency for a triple-buffered, five-stage pipe is thus ten time periods. In order to fill an n-stage pipe, $2n$ data packets must be input before any packets are output. Again, the overlap between input and output effectively keeps the third buffer from causing an extra delay.

Figure 4-12 shows the essential code for a triple-buffered pipeline. This code is similar to that of the double-buffered pipeline except that three pointers are needed to support three buffers, and the input, output, and processing all occur in parallel. The pointers are cycled around during each loop of the main process. The initial input of data to load the pipe is identical to that in Fig. 4-9.

Triple buffering a pipe makes maximal use of a processor's input and output capabilities but it does require triple the storage space of the simple, single-buffering method. Triple buffering is most useful when the processing to be done requires less time than the input plus the output. If this is not the case, double buffering will give equal or better performance, and with less storage overhead.

Pipeline Processing   *Chapter 4*

	T_0	T_1	T_2	T_3	T_4	T_5	T_6	T_7	T_8	T_9	T_{10}	T_{11}	T_{12}	T_{13}	T_{14}	T_{15}
	I_0	I_1	I_2	I_3	I_4	I_5	I_6	I_7	I_8	I_9	I_{10}	I_{11}	I_{12}	I_{13}	I_{14}	I_{15}
Processor 0		P_0	P_1	P_2	P_3	P_4	P_5	P_6	P_7	P_8	P_9	P_{10}	P_{11}	P_{12}	P_{13}	P_{14}
			O_0	O_1	O_2	O_3	O_4	O_5	O_6	O_7	O_8	O_9	O_{10}	O_{11}	O_{12}	O_{13}
			I_0	I_1	I_2	I_3	I_4	I_5	I_6	I_7	I_8	I_9	I_{10}	I_{11}	I_{12}	I_{13}
Processor 1				P_0	P_1	P_2	P_3	P_4	P_5	P_6	P_7	P_8	P_9	P_{10}	P_{11}	P_{12}
					O_0	O_1	O_2	O_3	O_4	O_5	O_6	O_7	O_8	O_9	O_{10}	O_{11}
					I_0	I_1	I_2	I_3	I_4	I_5	I_6	I_7	I_8	I_9	I_{10}	I_{11}
Processor 2						P_0	P_1	P_2	P_3	P_4	P_5	P_6	P_7	P_8	P_9	P_{10}
							O_0	O_1	O_2	O_3	O_4	O_5	O_6	O_7	O_8	O_9
							I_0	I_1	I_2	I_3	I_4	I_5	I_6	I_7	I_8	I_9
Processor 3								P_0	P_1	P_2	P_3	P_4	P_5	P_6	P_7	P_8
									O_0	O_1	O_2	O_3	O_4	O_5	O_6	O_7
									I_0	I_1	I_2	I_3	I_4	I_5	I_6	I_7
Processor 4										P_0	P_1	P_2	P_3	P_4	P_5	P_6
											O_0	O_1	O_2	O_3	O_4	O_5

Figure 4-11
Detailed activity diagram for a triple-buffered five-processor pipeline

```
VAL PacketSize IS 100:
PROC pipe(CHAN OF [PacketSize]INT left.in,right.out)
  ...  PROC work(data)
  [3][PacketSize]INT data:
  INT p0,p1,p2,temp:
  SEQ
    p0:=0                                       --initialize pointers
    p1:=1
    p2:=2
    [PacketSize]INT work.data  IS data[p1]:
    [PacketSize]INT out.data   IS data[p2]:
    [PacketSize]INT in.data    IS data[p0]:
    SEQ                                         --start pipe up
      left.in?   work.data                        --first input
      PAR
        left.in?  in.data                         --second input
        work(work.data)                           --first work
    WHILE TRUE
      SEQ                                       --main pipeline loop
        [PacketSize]INT work.data IS data[p0]:
        [PacketSize]INT out.data  IS data[p1]:
        [PacketSize]INT in.data   IS data[p2]:
        PAR
          left.in?   in.data                      --input
          work(work.data)                         --work
          right.out! out.data                     --output
        temp:=p0                                --exchange pointers
        p0:=p2
        p2:=p1
        p1:=temp
:
```

Figure 4-12
A pipeline processor with a triple buffer for work, input, and output

A Buffered Pipeline Test

To illustrate the advantages and disadvantages of each of the three buffering techniques, we measure the performance of each method on a 20-node pipeline processing 32768 data elements divided into 2048 packets and with various amounts of work per element. The packet size chosen is the most efficient packet size found in Fig. 4-5.

Figure 4-13 shows the results of the test comparing the efficiency of the three buffering techniques. The work per element ranges from zero microseconds (leaving only the communication overhead) to one thousand microseconds (used in Fig. 4-5). The measured results for each buffering method are shown in columns two

Pipeline Processing   Chapter 4

Work per element (µsecs)	Computation time (milliseconds) 32768 elements in 2048 packets, pipe of 20 processors			
	Single buffer	Double buffer	Triple buffer	Theoretical best
0	228	236	163	0
50	306	238	164	83
100	405	238	198	165
150	468	257	265	248
200	571	357	365	331
250	634	423	432	413
300	736	522	532	496
350	799	588	599	579
400	901	687	699	661
450	964	753	766	744
500	1067	853	866	827
1000	1890	1679	1700	1654

Figure 4-13
Tabulated performance of a 20-node pipe with different buffering schemes

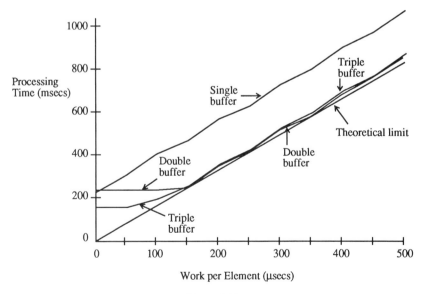

Figure 4-14
Performance plot of a 20-node pipe with different buffering schemes

to four; the best possible results, found from using Eq. 4-3, are shown in the final column.

The results of the tests using the three buffering methods are plotted in Fig. 4-14. As expected, single buffering is the slowest technique; this is not surprising since the input and output processes are run sequentially. Only when there is no work at all does single buffering have a very slight advantage over double buffering. This advantage is due to the double-buffer process overhead of switching between the parallel processes, exchanging pointers, and so on.

Double buffering basically uses the same sequential structure for the input and output as single buffering, and, as long as there is some work to do, double buffering is much superior to single buffering. The results using double buffering demonstrate that the communication is effectively performed in parallel with the computation. Until the work done on each element reaches about 150 μsecs, the communication takes longer than the computation and the pipe's performance remains constant. As the work done on each element increases past 150 μsecs, the work takes longer than the communication and the processing time increases with the work.

The efficiency results for the triple-buffered pipeline are similar to those for the double-buffered pipeline except that, since the input, output, and work are all overlapped, the performance of the triple-buffered processor pipeline is higher as long as the work takes less time than the input plus the output. Triple buffering does require a substantial amount of context switching to support each of the parallel processes, so that even with no work at all, a triple-buffered program will not execute twice as fast as a double-buffered program.

As the amount of work in a triple-buffered system increases beyond the time needed for the slowest data communication (either input or output), the processing time will begin to increase as the amount of work increases. When the processing time is greater than the time needed for both input and output, triple buffering produces the same results as double buffering. In fact, at that point, triple buffering is marginally worse than double buffering since it requires more overhead to organize three parallel processes.

The relative performance of the various buffering methods is summarized with the following observations, with P representing the time needed to process the work, I representing the time needed to do an input, and O the time needed to do an output:

- Double buffering is always superior to single buffering.
- A double-buffered system uses its resources most efficiently when $P = I + O$.
- The total computing time for a double-buffered system will be the maximum of the computing times for each buffer and will remain constant as long as the maximum does not change. If the input and output share one buffer and $P < I + O$, the total computing time will remain constant as P increases. When $P > I + O$, the total computing time will scale with P.

- A triple-buffered system uses its resources most efficiently when $P = I = O$.
- The total computing time for a triple-buffered system will be the maximum of the processing, input, and output times, and will remain constant as long as the maximum does not change.
- Triple buffering will be superior to double buffering when $P < I + O$ and $I < P + O$ and $O < P + I$. Again the last two cases are important only when the input and output times are not equal.

Multidimensional Pipelines

All of the pipeline examples discussed so far have had one dimension, and the data in the pipes have moved in one direction. With more than one bidirectional link connected to a processor, much richer pipeline structures can be built. For example, all of the previous pipelines mentioned in this chapter could have had two identical processes in every processor running totally separate pipelines in opposite directions (Fig. 4-15). This kind of structure is especially useful if the processing rate for a particular program is limited by the communication speed. Sending data through a pipe in both directions effectively doubles the communication bandwidth. But since there is still only one processor at each stage, the computing rate does not increase, and, if the computation at each node takes longer than half of the communication, there is little point in organizing a bidirectional data flow in a processor pipeline.

A more interesting situation arises if the data going in each direction is of a different type and the processors need both data flows to compute the results. This situation is quite possible in real-time systems which might obtain data from different sources and need to combine them in some fashion. Multidimensional pipelines provide a mechanism for implementing such systems.

Multidimensional pipelines can also be built quite easily. Figure 4-16 shows a two-dimensional structure with data flowing in two directions through a mesh of processors, from top to bottom and from left to right. Obviously, as is the case with any pipeline, the processors on the edge or end of an array must make special provision for the data input or output. All of the discussions and methods presented earlier for the one-dimensional pipeline are equally applicable to the multidimensional pipeline. In the two-dimensional example shown in Fig. 4-16, a triple-buff-

Figure 4-15
A pipeline of processors with data flowing in two directions

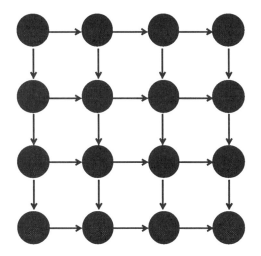

Figure 4-16
A two-dimensional pipeline of processors with data flowing in two directions

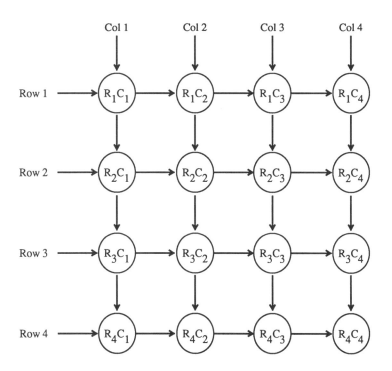

Figure 4-17
Matrix multiplication distributed over a two-dimensional pipeline

ered input/process/output structure can be readily designed for data that moves both vertically and horizontally.

Matrix multiplications are one commonly used example of a computation which can be done on a pipelined, two-dimensional grid of processors. If the rows of one matrix are passed through the pipe in one direction and the columns of the second matrix are passed through the pipe in the other direction, each processor can compute the portion of the result corresponding to the portions of the two matrices passed through it (Fig. 4-17). Each processor will compute a different portion of the final matrix since no two processors will receive both the same row of the first matrix and the same column of the second matrix. Notice that this case does not strictly meet our earlier definition of a pipeline system since not every data value passes through every processor. In a sense, the data in this situation is distributed over two sets of intersecting pipelines.

An example of the code used to implement a matrix multiplication in a single node is shown in Fig. 4-18. This program uses a single buffer for communication and, as a result, cannot overlap communication and computation. The program reads in NumBlocks of data in PacketSize groups. Data could be read in one element at a time but, as we have seen, it is more efficient to pass larger data sets with each communication. Each group then has its partial product computed and summed in the variable product.

The matrix multiplication program begins by zeroing product and entering a sequential loop with NumBlocks iterations. In each loop, the program simultaneously inputs a block of data from the left into row, and from above into col. After the data is read, it is passed to the right and down while the product of the new data elements is calculated and summed. When the loop is completed, all of the

```
PROC mult(CHAN OF [PacketSize]INT left.in,right.out,
                                  up.in,down.out)
  [PacketSize]INT row,col:
  INT product:
  SEQ
    product:=0
    SEQ i=0 FOR NumBlocks
      SEQ
        PAR                                                 --input
          left.in?   row
          up.in  ?   col
        PAR                                                 --output
          right.out! row
          down.out!  col
        SEQ i=0 FOR PacketSize                              --work
          product:=(row[i]*col[i])+product
:
```

Figure 4-18
Single-buffered node code for a matrix multiplication

data has passed through the node and the total product has been calculated. Notice that although the input cannot be overlapped with either the output or the product calculation, the output can be performed at the same time as the computation since the two operations do not assign any values to the same variables.

A much more complex way to do the same calculation is shown in Fig. 4-19. This routine is again an example of code in a single node which computes a matrix multiplication, but this time uses double buffering. Because the multiplication and sum will take more time than the communication of a data element, using triple buffering for the input and output is unlikely to give better performance.

In Fig. 4-19, two copies each of the arrays row and col are defined together with the pointers p0, p1, and temp. The variable product is again used to accumulate the product of the row and column element multiplications. Since the two buffers create a delay in the pipeline, the buffers must first be primed, then used alternately in a loop, and finally, emptied. To clarify the programming, we abbreviate both the row and col variables to work and io for their respective buffers. The io abbreviation is used for input and output while the work abbreviation is used to compute the product.

The buffer initialization is done after product is set to zero and p0 and p1 are set to zero and one respectively. After initialization, the work and io buffers are abbreviated. The first input from the left and top is read into work and the second input read into io while the first products from the work variable are summed in product. At this point the pipeline buffers are completely filled with the appropriate data.

The program then moves into its main loop. This loop is performed only NumBlocks−2 times to correct for the initial priming and final emptying of the buffers. In each iteration of the loop the work and io values are reabbreviated with the pointers. These pointers switch back and forth between the two row and col buffers. After the work and io buffers are selected, the data input and output begins with the io buffers while the product is calculated from the work buffers. The processor must first do an output to pass along the current values, and then an input to get the next data set. After the work and communication are done, the pointers are exchanged and the process repeats.

After all of the iterations in the main loop are complete, the pipe must be emptied. This procedure runs in reverse of the buffer initialization. The variable io is passed down and to the right while the product is accumulated with work. No data is read in. Finally, the last data is output and the program is finished.

While this example demonstrates a two-dimensional pipe with double-buffering, there is in fact no limit to the number of dimensions which can be used in this manner as long as the hardware can support the necessary communication. A three-dimensional structure, for example, only needs another link input or output command added in each of the input or output PAR structures and another dimension of links connected.

```
PROC mult(CHAN OF [PacketSize]INT left.in,right.out,
                                 up.in,down.out)
  [2][PacketSize]INT row,col:
  INT p0,p1,temp,product:
  SEQ
    product:=0
    p0:=0
    p1:=1
    [PacketSize]INT work.row IS row[p1]:
    [PacketSize]INT io.row   IS row[p0]:
    [PacketSize]INT work.col IS col[p1]:
    [PacketSize]INT io.col   IS col[p0]:
    SEQ
      PAR                                              --first input
        left.in?   work.row
        up.in  ?   work.col
      PAR
        PAR
          left.in?    io.row                           --second input
          up.in  ?    io.col
        SEQ i=0 FOR PacketSize                         --first work
          product:=(work.row[i]*work.col[i])+product
      SEQ j=0 FOR NumBlocks-2                   --main pipeline loop
        SEQ
          [PacketSize]INT work.row IS row[p0]:
          [PacketSize]INT io.row   IS row[p1]:
          [PacketSize]INT work.col IS col[p0]:
          [PacketSize]INT io.col   IS col[p1]:
          PAR
            PAR
              SEQ
                right.out!  io.row                     --pass row
                left.in?    io.row
              SEQ
                down.out!   io.col                     --pass col
                up.in  ?    io.col
            SEQ i=0 FOR PacketSize                     --work
              product:=(work.row[i]*work.col[i])+product
          temp:=p0                               --exchange pointers
          p0:=p1
          p1:=temp
```

Figure 4-19
Double-buffered node code for a matrix multiplication

```
    [PacketSize]INT work.row IS row[p0]:
    [PacketSize]INT io.row   IS row[p1]:
    [PacketSize]INT work.col IS col[p0]:
    [PacketSize]INT io.col   IS col[p1]:
    PAR                                             --empty pipe
      PAR
        right.out!   io.row                           --out io
        down.out!    io.col
      SEQ i=0 FOR PacketSize                        --work last
        product:=(work.row[i]*work.col[i])+product
    [PacketSize]INT io.row   IS row[p0]:
    [PacketSize]INT io.col   IS col[p0]:
    PAR                                             --out last
      right.out!   io.row
      down.out!    io.col
:
```

Figure 4-19 (cont.)
Double-buffered node code for a matrix multiplication

The different buffering techniques discussed earlier, single-, double-, and triple-buffering, can be applied equally well to higher-dimensional structures.

In Summary

Pipeline parallelism is a powerful computing method which is relatively easy to implement. Pipeline parallelism can provide performance flexibility with different pipe sizes and structures. Efficiently decomposing a problem to run on a pipeline is the most significant difficulty in using such an approach to parallelism. The overhead involved in communicating the data from one processor to another in the pipeline can also be significant.

Since all of the data processed in a pipeline typically passes through every processor, it is important to perform the data communication efficiently. The size of the data packets passed with each communication should be optimized for the pipeline length and the total amount of data. This communication can be done at the same time as the processing, but only at the expense of more buffers for storing data. Single buffering is most useful when the overhead of communication is less important than the cost of memory storage for data. Double buffering is most useful when the computation time equals the data input and output times together. Triple buffering is most useful for high-speed applications in which the data input, data output, and computation times are approximately equal, and the cost of data storage is small.

Pipeline structures are not limited to one dimension. Data can flow in many different ways depending on the configuration of the parallel computer. A bidirectional data flow can be used but will increase the communication bandwidth at the expense of greater storage requirements. Different data streams can also be merged in a processor where the different streams intersect. The same programs used to minimize the communication overhead for a one-dimensional data stream can be used in any multidimensional case.

Chapter 5

Data Parallelism

Data parallelism is the most common form of processing implemented on parallel computers. Computers using data parallelism generally distribute all of the data to be processed equally over all of the processors in the computer. Each processor is programmed to perform all of the processing on a subset of the data. This is in contrast to a pipeline system, in which the program is distributed rather than the data. Data parallelism is sometimes referred to as geometric parallelism.

Creating parallelism by distributing data is a popular approach because it largely avoids the difficulty of finding a way to decompose a problem into parallel pieces. In the distributed data approach, each processor contains the complete program. Not only is program decomposition largely irrelevant, but distributing data by dividing it equally among the processors also provides automatic load balancing. If data parallelism is to work efficiently, however, the task must have enough data with a small enough granularity to be sensibly divided. If there is not enough data or the granularity of the task is too large, some other approach is probably best.

At the same time that program decomposition becomes easier, however, interprocessor communication can become vastly more difficult. Because each processor contains only a portion of the entire data set, any processor which requires other, nonlocal data must obtain it by communicating with other processors. The difficulty of programming a processor network to communicate correctly and efficiently can more than compensate for the ease of program decomposition. Nevertheless, data parallelism is generally more flexible and easier to implement effectively than is pipeline parallelism.

In many ways, a multiple-instruction, multiple-data computer using data parallelism mimics the behavior of a single-instruction, multiple-data computer. Both machines perform the same operation with every processor at the same time, and both machines distribute data in much the same way. An MIMD computer implementing data parallelism does offer much greater communication flexibility than an SIMD computer, but at the expense of greater hardware cost.

Program Issues

Data parallelism is easily applied to a wide variety of architectures, including rings, toroids, and hypercubes. Generally, the architecture of the computing engine is chosen to match the computational requirements of the problems to be solved so as to minimize the communication overhead. For example, matrix operations are often performed on toroids or grids since the two-dimensional structure of a matrix is similar to a two-dimensional toroid or grid. Likewise, frequency transforms are effectively performed with hypercube architectures since the data as it is transformed never needs to move farther than one processor at each step. The actual distribution of the data in a parallel computer clearly depends upon the computer's architecture, but also upon the mathematical structure of the problem to be solved.

Data parallelism provides a high degree of flexibility within a parallel computer. Because the data is distributed over the processors, the computer's performance can scale (communication overhead permitting) directly with the amount of data or the number of processors. As the amount of data increases, each processor gets more work and simply slows down. If the amount of data decreases, performance improves. If the number of processors in a (regular) network increases, the amount of data in each processor decreases and the system performance improves. Note especially that these changes in amount of data and number of processors do not necessarily require any program modifications. In short, data parallelism provides flexibility by allowing the programmer or user to trade off processing speed with the number of processors and the amount of data processed. This flexibility is the fundamental appeal of all parallel processing, but it is most easily achieved with data parallelism.

Of course, it is simplistic to describe data parallelism by merely stating that each processor in a distributed system includes the entire program and a subset of the data. If indeed every data element can be processed independently of every other element, data parallelism *is* very simple. But very often the data elements cannot be processed alone; they must be combined in some fashion with other data. In this case, the additional data will sometimes be found in another processor, and some communication with the other processor will need to be done. This communication represents overhead which is not found in single-processor computers, and reduces the efficiency of computation.

The simplest distributed data architecture is a ring, shown in Fig. 5-1. The ring structure will be used for most of the examples in this chapter. Figure 5-2 illustrates the distribution of data and program tasks over a ring.

This chapter continues with a discussion of programming issues, followed by the presentation of a method for calculating the distribution of data over the processors. Three data loading techniques are presented with a comparison of the advantages and disadvantages of the different techniques. A routine for sampling a dis-

Figure 5-1
A ring network

Chapter 5 Data Parallelism

Figure 5-2
Task and data distribution for data parallelism

tributed data set is shown, after which two generic approaches for communicating nonlocal data are presented and compared.

Data Distribution

The first task to be done in any kind of data parallel system is to actually distribute the data. If the total number of data elements is a multiple of the number of processors, the data can be divided evenly among the processors. But more frequently the number of data elements is not a multiple of the number of processors, or, for some reason, the data cannot be evenly divided but must be organized in blocks which are not evenly divisible among the processors. This last situation could occur, for example, if each processor required an even number of data elements.

If the data cannot be evenly allocated among the processors, one or more processors will have more data, and therefore more work, than the others. This uneven distribution of data is another source of inefficiency in a parallel computer and is analogous to the uneven distribution of work in a pipeline system. In the most extreme case, it is possible that a processor might have no data at all. It is the programmer's task to ensure that this does not happen.

One simple approach to distributing data among a ring of four processors is illustrated in Fig. 5-3. In this example, the common data is the data which is evenly divisible over all of the processors. The remaining data which is not evenly divisible over the processors is distributed evenly over as many processors as possible, starting from the left processor. If the data must be distributed in blocks, any extra

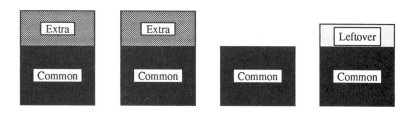

Figure 5-3
Distributing data over multiple processors

data less than a block in size will be left over and allocated to the last processor. Data can only be left over if the total amount of data is not evenly divisible by the block size. This happens, for example, when each processor must have an even number of data elements and the total amount of data is an odd number.

Figure 5-4 shows the code necessary for implementing this data distribution on the processor ring shown in Fig. 5-1. The first set of parameters passed to process setup describes the ring size and the amount of data. Parameter x.trans is the number of processors in the ring, and i.trans is the local processor number, which is calculated from a count of the processors from left to right around the ring, starting with zero. Value x.vec is the size of the vector of data, and x.block is the minimum-size block of data which can be distributed to a processor. Process setup returns three values: parameter i.vec is the number of elements in the local data partition, i.addr is the vector address of the first element in the local partition, and i.min is the smallest nonzero common data partition in any of the processors. This last parameter is useful when a processor must exchange data with a neighbor and needs to know the minimum amount of data the neighbor must have. Parameter i.max is the largest data partition found in any processor, and can be used in every processor to calculate the same data partition addresses.

The distribution calculation begins with a computation of the size of the common data partition. This size is simply the total amount of data divided by the number of processors and adjusted by the blocking factor. The remainder of data (x.rem) is then found, and also the number of odd, unallocated blocks of data (x.odd). The common data size is tested in the first IF structure and, if it is not zero, then i.min is set to the common data size plus, if all but the last processor get an extra block, any remainder which cannot be distributed in blocks. If the processor did calculate a common data partition size of zero, i.min is set to the leftover portion, or, if that is zero, to the block size. Parameter i.min is then the largest amount of data that every processor can be guaranteed to pass to a neighbor, assuming that processors with no data simply pass on data received from one neighbor to the next.

After the calculation of i.min, the routine continues with a second IF structure to adjust the maximum size value. If there are any extra blocks to be allocated (x.odd greater than zero), the maximum size of any processor's data partition (i.max) must be the common vector size plus the block size. If there are no extra blocks, the maximum local vector size is then the common size plus any remainder.

The address of the data partition (i.addr) is calculated next and initially set to the processor number times the common vector size. This calculation does not reflect any leftover data or extra blocks which might be stored in any processor to the left of the local processor. Therefore, if the processor address (i.trans) is less than the number of extra data blocks, the processor must store an extra block, and both i.vec and i.addr must be corrected to reflect the extra data. Any leftover data will be stored in the last processor and does not need to be included in the

```
PROC setup (VAL INT x.trans,i.trans,x.vec,x.block,
            INT i.vec,i.addr,i.min,i.max)
  INT x.rem,x.odd:
  SEQ
    i.vec := ((x.vec/x.block)/x.trans)*x.block    --common blocks
    x.rem  := x.vec-(i.vec*x.trans)               --left over blocks
    x.odd  := x.rem / x.block                     --left over elements

    IF                                            --set i.min
      i.vec <> 0                                  --normal case
        IF
          x.odd=(x.trans-1)                       --add left over
            i.min:=i.vec+(x.rem \ x.block)
          TRUE
            i.min:=i.vec
      TRUE                            --some processor has nothing
        SEQ
          i.min := (x.rem REM x.block)
          IF
            i.min=0
              i.min:=x.block
            TRUE
              SKIP

    IF                                            --set i.max
      x.odd > 0
        i.max:=i.vec+x.block
      TRUE
        i.max:=i.vec+x.rem
    i.addr := i.trans*i.vec                       --set i.addr

    IF                               --correct i.vec & i.addr
      i.trans < x.odd                             --extra blocks
        SEQ
          i.vec  := i.vec+x.block
          i.addr := i.addr+(x.block*i.trans)
      TRUE                                        --no extra blocks
        SEQ
          i.addr :=i.addr+(x.odd*x.block)
          IF
            i.trans = (x.trans-1)
              i.vec := i.vec+(x.rem REM x.block)
            TRUE
              SKIP
:
```

Figure 5-4
Code for calculating the distribution of code over a ring of processors

i.addr calculation. If the processor address is not less than the number of extra data blocks, the address must still be corrected for all of the extra data stored in processors with lower addresses. This correction is equal to x.odd times x.block; x.rem cannot be used since it also includes any leftover data.

At this point, only the data left over after the extra block correction remain. If the local processor is the last processor (i.trans equals x.trans minus one), it must add the leftover data to its data partition size. The amount of leftover data is the data remaining after the common data and the extra blocks (x.rem modulo x.block) have been subtracted.

As an example of this computation, consider a ring of four processors distributing a vector of 21 elements in blocks of two elements. Four is the largest multiple of two which, when multiplied by the number of processors (four), is less than 21. Therefore, at least four data elements are stored in every processor. This leaves five data elements to be distributed in blocks of two over the rest of the ring. Obviously, there are two blocks of two with a remainder of one left over to be distributed over the processors. The two lowest-address processors are given the extra blocks, and the single remaining value is assigned to the processor on the right. The result is illustrated in Fig. 5-5a, and the values of the various parameters are listed in Fig. 5-5b.

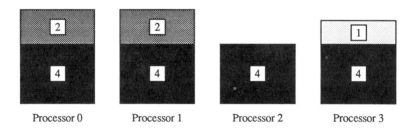

Figure 5-5a
Twenty-one data elements distributed in blocks of two over four processors

Processor	i.vec	i.addr	i.min	i.max
0	6	0	4	6
1	6	6	4	6
2	4	12	4	6
3	5	16	4	6

Figure 5-5b
Parameter values for example in Fig. 5-5a

This calculation of data distribution can be extended to multidimensional arrays of data. The distribution of data in each dimension is computed independently and is easily calculated using the same routine for each dimension.

Loading Data

Once the data distribution for a block of data has been calculated, the data must actually be loaded into the computer. Data loading is a source of communication overhead in a parallel computer, so it is important to minimize the time required for this operation, especially since many parallel computers cannot afford to connect every processor to an interface with the outer world. For processors without such an interface, the data must be passed through one processor to reach another. This requirement exacerbates the problem of communication overhead.

It is important to note that in the following examples, the data is assumed to distribute exactly evenly over the processors. With an exactly even distribution of data, the program code is simplest and clearest, but for any application in which processors might hold different amounts of data, the differences must be accommodated in the program. This accommodation is especially important for the routines using special buffering, since the data blocks may be of different sizes and thus require different-sized buffers.

A Simple Load Routine

Figure 5-6 lists a simple routine for distributing data geometrically over a ring. In the routine it is assumed that the right-most processor (with the highest address) reads in the data vector from the right and passes it to the left. The processor with the lowest processor address keeps the data with the lowest address; the processor with the highest address keeps the data with the highest address. As data is passed along, each processor inputs data in its turn, keeping whatever portion is allocated to it, and passing the rest to the left. The left-most processor will only receive data from the right and will never pass data onward. Eventually, the right-most processor will read in its portion of data and pass nothing on. The entire procedure could be run backwards, passing data from the left to the right and from the lowest address processor to the highest, if the data with the highest address were sent first.

The distribution process in the simple load routine uses the same parameters as the setup routine discussed earlier (Fig. 5-4). The program begins by calculating get, the count of data elements which it receives from the right. This value is the same as the i.addr value of the processor's neighbor to the right and is equal to the vector address of the first element plus the size of the local portion of the data. Four different situations are now possible. If get is zero, no data is expected and the routine is finished. If i.addr is zero, no data needs to be passed to the left and the routine simply reads in and stores the vector of data elements. If i.vec is zero, no data is stored locally and any data read in is immediately passed along

```
PROC load([]INT vector,buffer,VAL INT i.vec,i.addr,
    CHAN OF ANY left.in,left.out,right.in,right.out)
  VAL INT get IS i.addr+i.vec:
  IF
    get=0                                          --go to sleep
      SKIP
    i.addr=0                                       --input only
      right.in? [vector FROM 0 FOR get]
    i.vec=0                                        --pass only
      SEQ
        right.in? [buffer FROM 0 FOR get]
        left.out! [buffer FROM 0 FOR get]
    TRUE                                           --pass & input
      SEQ
        right.in? [buffer FROM 0 FOR get]
        PAR
          left.out! [buffer FROM 0 FOR i.addr]
          [vector FROM 0 FOR i.vec]:=
                        [buffer FROM i.addr FOR i.vec]
:
```

Figure 5-6
Simple code for data distribution on a ring

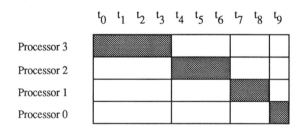

Figure 5-7
Link activity diagram for the simple load routine

to the left. Finally, if none of the previous cases is true, the routine must read in the data elements, store some of them locally in vector, and pass the rest along.

The data transfers through the links are done in this example with simple block input and output commands, but they can also be done with counted string protocols. However, it is not necessary to use protocols since every processor can calculate the amount of data to input, output, and keep.

This loading routine is simple, but because only one link is transferring data at a time, it is inefficient, especially for large amounts of data. Figure 5-7 is a four-processor activity diagram of the link communications for each processor. A filled box represents a busy link loading data into the processor. Four time units are needed to load all of the data into the first processor, one time unit for each of the four data partitions. The second transfer only requires three time blocks since the first processor keeps one data partition. The third transfer requires two time blocks, and the final transfer moves the last data partition into the last processor. In all, 10 time blocks are needed to complete the load. Since only 10 of the 40 available communication time blocks are used, the communication efficiency is 25%.

A Fast Load Routine

The rate at which data are loaded into a processor network can be substantially improved by the use of some pipelining techniques. If each processor double buffers the data input and output so that both input and output can occur simultaneously, the loading performance is much better. Figure 5-8 is an activity diagram for such a pipeline load routine. First, the data partition for processor 0 is loaded into processor 3. As the second partition is loaded, the first one is passed to processor 2. This procedure continues until the last data partition is loaded into processor 0, concurrently with every processor receiving its correct data set. The total loading time is now four time blocks. Since 10 of the 16 time blocks are used for data movement in this second example, the communication efficiency is 62.5%. As the ring network grows larger, this efficiency will asymptotically approach 50%.

The code necessary for implementing the pipeline loading technique is strongly reminiscent of the code used for pipeline communication in Chap. 4, and is shown in Fig. 5-9. In this routine, data is moved in exactly the same way as in

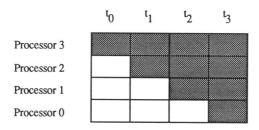

Figure 5-8
Link activity diagram for the fast load routine

```
PROC load2([]INT vector,
          VAL INT i.vec,i.addr,
        CHAN OF ANY left.in,left.out,right.in,right.out)
  [2][block.size]INT buffer:
  IF
    (i.addr+i.vec) = 0                      --nothing to read
      SKIP
    i.addr=0                                --nothing to pass
      VAL INT number.of.blocks IS i.vec/block.size:
      SEQ i=0 FOR number.of.blocks
        right.in?  [vector FROM (i*block.size)
                           FOR block.size]
    i.vec=0                                 --nothing to keep
      INT pointa,pointb,temp:
      VAL INT number.of.blocks IS i.addr/block.size:
      SEQ
        pointa:=0                           --initialize pointers
        pointb:=1
        right.in? buffer[1]                 --first input
        SEQ i=0 FOR (number.of.blocks-1)    --main loop
          SEQ
            buffer.in  IS buffer[pointa]:
            buffer.out IS buffer[pointb]:
            PAR                             --in & out
              right.in? buffer.in
              left.out! buffer.out
            temp:=pointa                    --exchange pointers
            pointa:=pointb
            pointb:=temp
        left.out! buffer[pointb]            --empty buffer
    TRUE                                    --read keep and pass
      VAL INT number.of.blocks IS i.addr/block.size:
      INT pointa,pointb,temp:
      SEQ
        SEQ
          pointa:=0                         --initialize pointers
          pointb:=1
          right.in? buffer[1]               --first input
          SEQ i=0 FOR (number.of.blocks-1)  --main loop
            SEQ
              buffer.in  IS buffer[pointa]:
              buffer.out IS buffer[pointb]:
              PAR                           --in & out
                right.in? buffer.in
                left.out! buffer.out
              temp:=pointa                  --exchange pointers
              pointa:=pointb
              pointb:=temp
```

Figure 5-9
Code listing for the fast load routine

```
    PAR
      left.out! buffer[pointb]                    --empty buffer
      VAL INT number.of.blocks IS i.vec/block.size:
      SEQ i=0 FOR number.of.blocks               --store last
        right.in? [vector FROM (i*block.size)
                          FOR block.size]
:
```

Figure 5-9 (cont.)
Code listing for the fast load routine

the double-buffered pipeline. Two buffers are defined, and as one is filled the other is emptied; the pointers to each buffer are exchanged so that what was an input buffer becomes an output and vice versa; and then the procedure repeats. Note that the last data communication will occur without a corresponding input, and that each processor will simultaneously input its final data set.

The code for the fast load routine is organized so that the buffers can be of any size that evenly divides the size of the data partition. For the case diagrammed in Fig. 5-8, the buffer is the same size as the data partition. It is possible to use smaller buffers, thereby saving memory space. Figure 5-10 is an activity diagram showing the effect of using buffers one-half the size of each data partition. When half-sized buffers are used, the communication of each data set must be done in two stages. Although twice as many time blocks are used in the half-size buffer case, each time block is one-half the size of the time block in the full-size buffer case, since half as large a data block is being moved with each communication. This shows that the buffer size itself makes no difference in the program's efficiency or performance. For this case, 20 of 32 time blocks are used, resulting in a communication efficiency rate of 62.5%, which is exactly the same as the rate achieved using a full-size buffer. However, each data communication *does* involve some organizational overhead which increases the program overhead. Since the half-size buffer case requires twice as many communications as the full-size case, it will have a

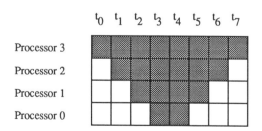

Figure 5-10
Link activity diagram for the fast load routine with half-size buffers

Data Parallelism   Chapter 5

slightly lower performance. This lower performance must be traded off against the cost of larger memory buffers.

Bidirectional Loading

One of the very useful features of the transputer links is their bidirectional capability, which means that data can be passed around a ring in two directions at the same time. In the previous example, data was only passed in one direction. By passing data in both directions, we should be able to double the performance of the loading routine.

The first question we face in considering bidirectional loading is whether data should be passed all of the way around the ring in both directions, or only halfway around in each direction. Figure 5-11 is an activity diagram for the case in which data is passed only halfway around the ring in each direction. In this case, only two blocks of data are passed into the ring in each direction, the total time required is two time blocks, and the total link utilization is three time blocks out of eight, for a communication efficiency rate of 37.5%.

If, instead of passing data only halfway around the ring, we pass data all of the way around in both directions, the efficiency is higher, as shown in Fig. 5-12. Here, each link is passing half as much data in half the time on each transfer, but the transfers continue all of the way around the ring in both directions. The efficiency is now back to 62.5%, but the same amount of time is required to actually complete the load. The data itself must travel further, so the higher link utilization does not improve the overall performance. We can prove this by noting that the half-way routine moves six data blocks one processor, while the full-way load moves ten data blocks one processor. Six blocks divided by 37.5% efficiency equals ten blocks divided by 62.5% efficiency.

Although both of these approaches have the same overall performance, their link utilization and data distribution differ. For both loading routines, the actual bandwidth constraint is the rate at which data can be moved out of the loading processor.

Both of these bidirectional load routines can be implemented with the pipeline load routine listed in Fig. 5-9. The halfway-around routine (Fig. 5-13) is essentially a one-direction load. Half of the processors will call the load routine so as to pass data in one direction, and the other half will pass data in the other. Each processor tests itself to see which side of the ring it inhabits (with respect to the processor from which the data is coming), and then "pretends" to be part of a linear array whose end is the processor from which the data is coming. The routine listed in Fig. 5-13 does this by testing `i.trans` and then calling `setup` with the appropriately adjusted parameters.

The bidirectional load which passes data all of the way around the ring uses two parallel calls to the pipeline load routine (Fig. 5-14). Half of the data is passed each way, and the two load calls order their links oppositely and use separate vector storage variables.

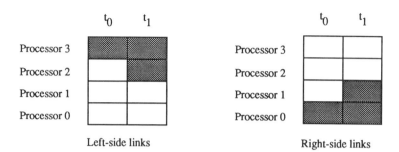

Figure 5-11
Link activity diagram for a bidirectional load going halfway around the ring

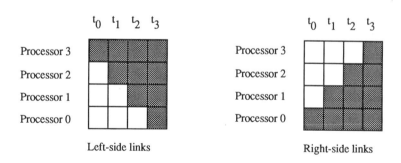

Figure 5-12
Link activity diagram for a bidirectional load going completely around the ring

```
IF
  i.trans >= (x.trans/2)                              --on right side
    VAL h IS x.trans/2:
    SEQ
      setup (x.trans,i.trans-h,x.vec,x.block,
                       i.vec,i.addr,i.min,i.max)
      load2(vector,i.vec,i.addr,
             left.in,left.out,right.in,right.out)
  TRUE                                                --on left side
    VAL h IS (x.trans/2)-1:
    SEQ
      setup (x.trans,hi.trans,x.vec,x.block,
                       i.vec,i.addr,i.min,i.max)
      load2(vector,i.vec,i.addr,
             right.in,right.out,left.in,left.out)
```

Figure 5-13
Code for initiating a bidirectional, halfway load routine

```
PAR
  load2(vector.a,i.vec,i.addr,                        --pass to the left
        left.in,left.out,right.in,right.out)
  load2(vector.b,i.vec,(x.vec-i.vec)-i.addr,          --pass to the right
        right.in,right.out,left.in,left.out)
```

Figure 5-14
Code for initiating a bidirectional, full-way load routine

Figure 5-15
Necessary distribution of data packets for the halfway load routine

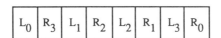

Figure 5-16
Necessary distribution of data packets for the full-way load routine

While the bidirectional loading routines clearly give the best performance of the different routines discussed, there is an important issue to consider concerning the data's original organization in memory. If we use a bidirectional routine, the data must be loaded from noncontiguous memory blocks. If one final block with data stored in address order is desired, the processor loading the ring must be careful to pass the correct portion of the data in the correct direction. Figure 5-15 shows the necessary organization of the data vector in the loading processor for the halfway loading routine; Fig. 5-16 shows the necessary organization for the full-way routine. In both figures, L indicates a packet which must be sent to the left, R indicates a packet to be sent to the right, and the subscript value shows the order in which the packets must be sent.

Since both of these routines have the same performance, issues of data organization or number of processors may well be the deciding factor in any choice. The halfway around routine does have one important advantage. Since it only uses links in one direction, an unload operation can be done at the same time by using the links to communicate data in the opposite direction. This application combines the greatest possible use of the links with the minimum data communication distance. However, if the ring size is odd, note that the halfway routine will either become less efficient or more complicated to handle the case of the extra processor.

Performance Measures

The performance of each of these five basic routines for a 12-processor ring loading 98304 32-bit values is shown in Fig. 5-17. The first column reports the measured performance and the second column gives the theoretically best performance. The theoretically best performance is extrapolated from the timing of the simple load program using the efficiency calculations done earlier. Each processor will load 8192 values, evenly distributing the entire data set. The simple load routine requires 78 transfers of 8192 words (12 + 11 + 10 + ... 2 + 1) to completely distribute the data. This is done in 1.53 seconds for an overall data rate of 1.67 MBytes/sec. Because it uses double buffering to overlap input and output, the fast

Loading routine	Measured	Best	Efficiency
Simple load	1.53 secs	1.53 secs	-----
Fast load with 8k-word buffer	0.29 secs	0.24 secs	83%
Fast load with 32-word buffer	0.31 secs	0.24 secs	77%
Bidirectional load (halfway) with 4k buffer	0.15 secs	0.12 secs	80%
Bidirectional load (full-way) with 4k buffer	0.15 secs	0.12 secs	80%

Figure 5-17
Measured and theoretically best performance for loading routines

load routine requires only 12 time blocks to complete the transfers needed to load the data. The best performance that could be expected using the fast load routine would be 12 / 78 of the simple load time. Using much smaller buffers, the fast load routine could theoretically load data at the same rate, but the increased communication overhead for the additional transfers does reduce the overall performance. The bidirectional load (using the largest possible buffers) should require one-half the fast load time, but again, increased communication overhead slightly reduces the overall performance.

As judged from the measured performance figures, these simple loading routines are reasonably effective. The bidirectional load achieves an efficiency of 80% (assuming the simple load performance as a base for extrapolation); the other routines are roughly as effective. Compared with the fast load routine, the bidirectional loads are nearly 97% efficient. The fast load with 32-word buffers is somewhat less efficient at 77%.

Sampling

Once a data set has been loaded, it can be manipulated in parallel by the many processors in a network. Sometimes, however, only sampled portions of a data set need to be processed. Operations are often performed on a sampled subset of data, and when the data is distributed over many processors it may not be obvious which data elements in each processor are members of the subset. If a set of data is distributed using the program listed in Fig. 5-4, the location of any subset of the data can be determined in each processor using the code listed in Fig. 5-18.

The parameters passed to the sample.calc routine define the sampling of the distributed data set. Five values are passed to the routine: first, the size of the local portion of the data vector (i.vec); second, the address of the first element in that portion (i.addr); third, the starting element of the subset (start); fourth, the ending element of the subset (end); and fifth, the sampling frequency of the subset (skip). These last three values are the sampling values and are passed to all of the processors. The first value (start) is the address of the first value in the sampled subset, the second value (end) is the address of the last value in the subset, and the third value (skip) is the sampling period. Calculated from this information are the offset address (from the beginning of the local data set) of the first subset data element in each processor (offset), the total number of subset elements stored in processors with a lower address (before), the total number of subset elements stored in the processor itself (local), and the total number of subset elements stored in processors after it (after).

The sampling calculation listed in Fig. 5-18 begins with the clearing of the values to be returned. The local value r.vec, representing the number of elements in the sample, is calculated. The address of the local portion of the data set is then checked to see if any of the sampled set is held locally. If the entire sample is stored at an address greater than that of the local partition, after is set equal to r.vec. If the ending address of the data sample is less than the address of the local parti-

```
PROC sample.calc( INT offset,before,local,after,
            VAL INT i.vec,i.addr,start,end,skip)
  INT r.vec:
  SEQ
    offset:=0
    before:=0
    after :=0
    local :=0
    r.vec:=((end-start)/skip)+1               --vector size
    IF
      start >= (i.addr+i.vec)                 --starts after me
        after := r.vec
      end < i.addr                            --ends before me
        before := r.vec
      TRUE                                    --overlaps me
        INT r,q:
        SEQ
          q    := ((i.addr+skip)-1)-start
          r    := (q+i.vec)/skip
          q    := q/skip
          IF                                  --find before me
            start < i.addr
              before := q
            TRUE
              SKIP
          IF                                  --find after me
            end >= (i.addr+i.vec)
              after := r.vec-r
            TRUE
              SKIP
          local:=r.vec-(before+after)         --find in me
          IF
            (local>0)
              offset:=((before*skip)+start)-i.addr
            TRUE
              SKIP
:
```

Figure 5-18
Calculation for sampling a data set

tion, `before` is set equal to `r.vec`. If neither of these two cases is true, the data set overlaps the local processor itself, which may then contain some of the sampled set. The number of sampled elements stored before the first local element is then calculated and assigned to `q`, and the number of values stored after the local partition and corrected for the sampling frequency is assigned to `r`. After being set once, `q` is corrected for the sampling frequency. If the sampled data begin before the local partition, `before` is set to `q`. If the sampled data end after the local partition, `after` is set to the sampled vector size `r`. The local portion of the sampled data set must then be the sampled vector size (`r.vec`) minus the `before` and `after` portions. If `local` is greater than zero, the address offset of the first sampled data element in the local data partition is simply the difference between the address of the first local sampled element and the `i.addr` value. The address of the first local sampled element is the sum of the `start` value and the product of the `skip` and `before` values.

Just as the original distribution calculation is independent of dimension, so is this sampling calculation. To calculate the distribution of a sampled data set, this routine can be called as many times as there are dimensions in the particular processor architecture.

Expanding Data Sets

Data parallelism is extremely efficient as long as all data can be completely processed locally inside a processor. This usually happens when data elements are processed individually, without regard to any other element. There are many operations, however, which do combine different data elements. In this case, each processor must get information from neighboring processors to perform the operation.

There are many ways in which a parallel, distributed computer can move data internally to support nonlocal operations. Two different approaches are discussed here, but these approaches are not mutually exclusive, nor are they the only possibilities. In the first approach, the local data set is expanded so that redundant information is stored in each processor and does not need to be communicated. In the second approach, no extra data is stored locally, but data is passed from processor to processor so that each processor can temporarily take whatever data it needs to perform its operation. Both approaches obviously require extra communication, and so the overall efficiency of the parallel computer will suffer.

A convolution is a simple and typical operation for which each element requires information from neighboring elements. In a convolution, each data element in an array is replaced with the sum of itself and the neighboring elements multiplied by a set of constants called the kernel. The more values there are in the kernel, the more neighboring data elements are needed for the results to be calculated. Notice that some special provision must be made for data elements at the edges of an array. That issue is ignored here. On a closed surface such as a toroid, the data could simply be wrapped around the array, or special code written to handle the special case.

One-Dimensional Expansion

In order for a system to perform a convolution on a distributed data set, each processor must exchange data with its neighbor. Figure 5-19 illustrates one way of performing the exchange of information for a one-dimensional set of data distributed over a ring. To begin with, each processor has a subset of the data organized as an array in memory. These data are then copied into the middle of a larger memory storage area. Next, the data set in each processor is expanded, or enlarged, so that the data on the edges of each local subarray are augmented by the contiguous data held in the neighboring processor. The data on the edges are now held in both processors, and each processor has all of the data necessary to perform the convolution.

A simple code fragment which can accomplish this exchange of data is listed in Fig. 5-20. The routine first moves data from the src buffer to the middle of the expand buffer. The values left and right describe the overlap to be achieved on the left and right sides of the array of data elements respectively, and must be large enough to accommodate all of the data necessary for the convolution of the first and last data elements in the processor's array. Four abbreviations are then created to exchange the correct portions of the data: l.pass is an abbreviation for the data sent to the left, r.pass for the data sent to the right, l.read for the data input from the left, and r.read for the data input from the right. The PAR structure then performs both exchanges in parallel.

Two-Dimensional Expansion

The one-dimensional case is quite simple compared to the complexities encountered in the two-dimensional case. If an array of values is distributed over a two-dimensional grid, or toroid, of processors, each processor will have a two-dimensional subarray of the data. If a two-dimensional convolution is performed on the array, the overlap between subarrays must also be two-dimensional so that a rectangular array of original data is stored within a larger rectangular array of expanded data. Figure 5-21 illustrates the exchanges necessary for expanding the subarray in two dimensions. The white portion of the box shows the original data with the left, right, top, and bottom portions to be exchanged marked off by dotted lines. On the left-hand portion of the figure, the arrows indicate the exchange of data to the left and right. The overlapped data for the left-right direction are stored in the lightly dotted region. On the right-hand portion of the figure, the arrows indicate the subsequent exchange of data for the top and bottom. The overlapped data for the top-bottom direction are stored in the more heavily dotted region.

Notice that for a rectangular grid, this expansion must be a two-step process since each processor needs data from the diagonally neighboring processors. Data from the right and left neighbors are exchanged and then passed to the top and bottom processors in the second step; this two-step process provides the data for the corners of the larger, expanded array. If each processor is connected to its eight nearest grid neighbors, then all eight exchanges can be done in parallel.

Data Parallelism   Chapter 5

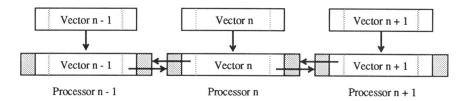

Figure 5-19
A data set distributed and expanded over multiple processors

```
PROC expand([]INT32 src,expand,
            VAL INT left,right,i.vec,
            CHAN OF ANY left.in,left.out,right.in,right.out)
  SEQ
    [expand FROM left FOR i.vec-(left+right)]:=src
    l.pass IS [expand FROM left      FOR left]:
    l.read IS [expand FROM 0         FOR left]:
    r.pass IS [expand FROM i.vec-(right*2) FOR right]:
    r.read IS [expand FROM i.vec-right     FOR right]:
    PAR
      left.out ! l.pass
      left.in  ? l.read
      right.out! r.pass
      right.in ? r.read
:
```

Figure 5-20
Code listing for a one-dimensional expansion

132   Parallel Programs for the Transputer

Chapter 5   Data Parallelism

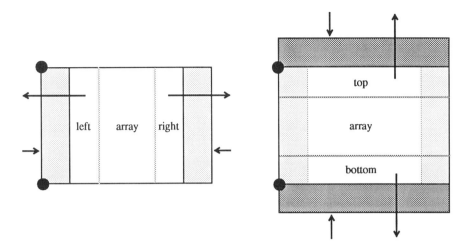

Figure 5-21
A two-dimensional expansion

This two-dimensional expansion is more difficult than the one-dimensional expansion because of the way data is stored in the computer. Memory is addressed as a one-dimensional array, and it is very straightforward to do an input or an output on a sequential, contiguous block of data stored in memory as in the one-dimensional expansion (Fig. 5-19). In the two-dimensional expansion (Fig. 5-21), the left and right columns of data are not stored contiguously in memory and therefore cannot be communicated as a block. The data must either be passed in small, one-dimensional subblocks or copied into a buffer which can then be communicated with one command.

The example in Fig. 5-22 demonstrates a two-dimensional expansion using the first of these two approaches. This example uses a single, generic exchange routine (swap) to perform the data exchanges. The swap routine is called four times to do the exchanges in each of the four directions. The expand routine is called with a set of arguments: array is the expanded array; i.array and j.array describe the width and height of the array; si.array and sj.array are the width and height of the original, unexpanded data set; local.size is the size of the expanded array; and left, right, top, and bottom define the overlap in each of those directions, so that i.array = si.array + left + right and j.array = sj.array + top + bottom. The eight communication channels for doing input and output in each of the four directions are included last.

The swap routine actually performs the exchanges. It takes as arguments repeat (the number of times data is actually to be exchanged), block (the amount of data to be exchanged), the start addresses for the source and destination of the data, and an input and output channel pair.

Parallel Programs for the Transputer   *133*

```
PROC expand.2d([]INT array,                          --subroutine
        VAL INT i.array,j.array,sj.array,local.size,
                left,right,top,bottom,
     CHAN OF INT left.in,left.out,right.in,right.out,
                up.in,up.out,down.in,down.out)
  INT src.point,dst.point:
  PROC swap(VAL INT repeat,block,start.s,start.d,
            CHAN OF INT out,in)
    INT src.point,dst.point:
    IF
      block>0
        IF
          start.s=start.d         --I have no data, pass through
            data IS [array FROM start.d FOR block]:
            SEQ q=0 FOR repeat
              SEQ
                in?  data
                out! data
          TRUE                                      --I have data
            SEQ
              src.point:=start.s                    --setup ptrs
              dst.point:=start.d
              SEQ q=0 FOR repeat
                SEQ
                  array.out IS  [array FROM src.point
                                       FOR block]:
                  array.in  IS  [array FROM dst.point
                                       FOR block]:
                  PAR                               --exchange
                    out! array.out
                    in?  array.in
                  dst.point:=dst.point+i.array      --update ptrs
                  src.point:=src.point+i.array
      TRUE
        SKIP
  :
  VAL start.a IS (top*i.array):                     --main routine
  VAL start.b IS (local.size-(bottom*i.array)):
  SEQ
    swap(sj.array,right,start.a+left,               --send left
                  start.a+(i.array-right),
                  left.out,right.in)
    swap(sj.array,left,(start.a+i.array)-(left+right),
                  start.a,right.out,left.in)        --send right
    swap(1,i.array*bottom,                          --send up
          start.a,start.b,  up.out,down.in)
    swap(1,i.array*top,                             --send down
          start.b-start.a,0,down.out,up.in)
:
```

Figure 5-22
Calculation for sampling a data set

If the amount of data to be expanded in a processor is zero, the whole routine is abandoned. If there is some data to be exchanged, it is still possible that the local processor itself does not have any and must merely pass data from one processor to another. When this is the case, the routine simply reads data from the input channel and sends the data to the output channel `repeat` number of times. If, as is normally the case, the local processor does have data, they must be exchanged. Two pointers (`src.point` and `dst.point`) are initialized to the source and destination addresses, and the exchanges begin. For `repeat` cycles, the routine does a parallel input and output of a block of values starting at the pointers. After each exchange, the pointers are updated and the procedure repeats.

The main program in Fig. 5-22 assumes that the original array has already been copied into the expanded array buffer, and begins by initializing two pointers, `start.a` (indicated by the upper dot in Fig. 5-21) and `start.b` (the lower dot in Fig. 5-21). The program proceeds by calling `swap` to do an exchange, passing data to the left and receiving data from the right, and filling the buffer on the right. The program repeats the exchange for each line in the original, unexpanded array by passing `right` values starting from the first element of the original array, and reading them in starting at the last element of the first line of the original array. After each exchange, the address pointers in `swap` are updated by one full line of the expanded array so that the entire block is passed sequentially, one line segment at a time. The second exchange on the left side is done in the same way but with the pointers reversed. The source of data is `left` elements left of the first element of the original array, and the destination is simply `start.a`.

The up and down transfers in Fig. 5-22 are considerably easier to perform than the left and right, since the data can be moved in one block. The repeat value is set to one, the block sizes are `i.array` times the `top` or `bottom`, and the source and destination pointers simply use `start.a` and `start.b`.

In this expansion routine, the right and left (or up and down) exchanges could actually be done in parallel, since the destination buffers do not overlap. However, if the right or left values are large enough, it is possible that the source blocks overlap. This makes it difficult for the programmer to use abbreviations or to completely check the correctness of the program in occam, and so the exchanges are done sequentially.

The expansion routine considered also does not take into account exchanged buffers which might be larger than the portion of the array stored locally. For example, the size of the local array might be 10 elements by 10 elements, while the overlap might be 50 by 50. In that case, the expansion operation must be done repeatedly. This situation will occur whenever `si.array < left` or `si.array < right` or `sj.array < top` or `sj.array < bottom`.

Performance Issues

The overhead involved in expanding a local data set is relatively small as long as the amount of data expanded is small relative to the total amount of pro-

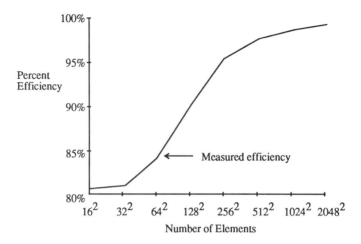

Figure 5-23
Computational efficiency for various data set sizes

cessing done on the data locally. If the data set is large and the expansion is small, efficiency will be high. Likewise, if the amount of processing per element takes a long time relative to the time required to get nonlocal information, efficiency will be high.

The relative efficiency of a ten-node by twelve-node toroid processing data sets of different sizes is shown in Fig. 5-23. We test for efficiency by expanding a data set and performing a nine-element two-dimensional convolution. The time required to complete the test is measured and compared to the time required to simply perform the convolution without exchanging the edge data. The data sets tested range in size from a 16-by-16 array to a 2048-by-2048 array. With the largest data set, the efficiency approaches 99%, while the smallest set which can be distributed over a 120-processor array has an efficiency of just over 80%.

Notice that the amount of data exchanged depends upon the number of data elements on the edge of each processor. If a processor contains a 100-by-100 array of data elements, it must first be augmented by 100 values on each of the first two sides, and then by 102 values on each of the two remaining sides (including the two new rows or columns from the first exchange), for a total of 404 values. If the array quadruples in size to 200-by-200, the number of elements communicated does not quite double, to 804 values. Thus the amount of communication needed to perform the same task (a nine-point convolution) decreases relative to the amount of work as the size of the data set increases. The amount of work varies with the number of data elements while the communication varies with the square root of this number.

While expanding data sets to get information from other processors is reasonably efficient, this approach does require large amounts of memory. If larger and larger overlaps are needed for a computation, the processors may eventually reach the natural limit at which every processor stores the entire data set from all of the processors. While this approach may be reasonable for some problems, the storage cost is very high.

Communicating Data Sets

Expanding a local data set is not the only way to provide nonlocal data to a processor. A more common approach is to write communication routines that simply pass the data from processor to processor until each processor has received whatever data it needs. Most of the communicated data does not need to be stored permanently; thus the memory requirements within the system are not as great. Matrix transpositions or multiplications, frequency transformations, and sorting programs are examples of routines that are often programmed using communication routines and do not store redundant data.

Shifting

As an illustration of one simple communication routine, consider once again a convolution on an array of data distributed over a toroid of processors. In the previous section we demonstrated how we can perform a convolution on a distributed set of data by enlarging the local data store so that each processor contains redundant information. We can also do this operation by shifting an entire copy of the data array and combining the shifted array with the original. This second approach requires less storage than the expanded data set approach since the processors do not need to store any redundant data. However, a separate shift of the data set is required so that the data set for each nonzero element of the kernel is aligned.

Figure 5-24 illustrates this shifting approach to data communication with an upward shift of data for a two-dimensional data set. The data at the top of the array is passed upward to the processor above the local processor, the main body of the data is moved internally, and a new set of data is received from the processor below the local one, and stored at the bottom of the array. A downward shift does the same in reverse. A shift to the left or right is conceptually the same, but, because the data is not stored in one sequential block, the transfers must either be done one line at a time, or the data must be reassembled in a contiguous memory block and passed as a group.

A program which implements an upward or downward shift of a distributed data array is listed in Figs. 5-25 and 5-26. Figure 5-25 is a subroutine which actually does the exchange of data between processors, while Fig. 5-26 is the controlling routine.

Process move in Fig. 5-25 is called with an array argument containing the data to be shifted (array), a temporary storage buffer, the size of the block to be moved (blk.siz), a pointer to the start of the data to be passed (start.-send), a pointer to the start of the storage location of the data to be received

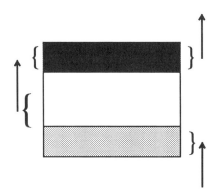

Figure 5-24
Shifting the local piece of a distributed data set up

```
PROC move([]INT array,buffer,
      VAL INT blk.siz,start.send,start.rec,mov.dir,size,
    CHAN OF ANY in,out)
  SEQ
    PAR
      in?  [buffer FROM 0 FOR blk.siz]
      out! [array FROM start.send FOR blk.siz]
    VAL INT amt.to.move IS size-blk.siz:
    INT k:
    IF
      amt.to.move<=0
        SKIP
      mov.dir>0
        SEQ
          k:=0
          WHILE (k<amt.to.move)
            SEQ
              array[k]:=array[k+blk.siz]
              k:=k+1
      TRUE
        SEQ
          k:=size
          WHILE (k>blk.siz)
            SEQ
              k:=k-1
              array[k]:=array[k-blk.siz]
    [array FROM start.rec FOR blk.siz]:=
                             [buffer FROM 0 FOR blk.siz]
:
```

Figure 5-25
Code for shifting a data set upward or downward

(start.rec), an argument indicating the direction of the move (mov.dir), the size of the local data array, and input and output channels.

The move process for shifting data begins with a parallel output of the data to be passed, and an input of the new data into buffer. Next, the new data must be stored and the remaining local data relocated correctly. The remaining local data to be shifted (amt.to.mov) is clearly the difference between the original array size and the amount of data already passed. If the move direction is greater than zero, the local data must be moved to a lower address. This is done in a simple WHILE loop, one element at a time beginning at element zero. If the move direction is less than zero, the local data is moved to a higher address, again with a simple WHILE loop, but with the array subscript value beginning at the last element and being decremented instead of incremented. The data can also be moved in a series of block memory assignments, each smaller than the block size. After the local data has been shifted, the new data received from the neighboring processor is assigned to the correct spot in the local array.

This move routine can shift a data set to the left or right, as well as up or down, if it is called a number of times to pass a subset of a single line. Of course, the pointers into the array must then be assigned and updated differently than when a single block is moved up or down.

The data movement is controlled with calls to process move. The process supports data shifts of arbitrary size (even shifts greater than the size of the local data in a processor) and can pass data either upward or downward. A similar but somewhat more complex routine can move data to the left or right.

The shifting routine in Fig. 5-26 is called with the same array and buffer arguments as the move subroutine, with the addition of the local.size of the data array, the minimum size of any processor's local data array in the x and y dimensions, and the number of lines to shift (positive for upward, negative for downward).

The size of the data block to be passed from processor to processor is the first value calculated by the shifting routine. This blk.mov is the minimum data size that any processor must contain. Remember that every processor in a column will have the same local line length for the local data array so that blk.mov will be identical for every processor in a column. After the size of the data block has been calculated, the processors must transfer the appropriate amount of data in blocks of this size. The total amount of data to be passed is found in shift.blk.

The shift routine continues by testing the direction of the shift. If the shift value is less than zero, data will be passed down. The number of blocks.to.- move is calculated and the move subroutine called that number of times. The size of the block to shift is blk.mov, and the pointer to the data to be passed is set to the local size minus the block size. Data is received at the first element of the array.

Once all of the complete blocks of data have been passed, the routine continues by passing whatever data remains. The value of left.over is simply the amount of data remaining after all of the complete blocks of data have been moved.

Data Parallelism

```
PROC shifty([]INT array,buffer,
         VAL INT local.size,i.array,i.min,j.min,shift,
      CHAN OF ANY up.in,up.out,down.in,down.out)
  VAL INT shift.blk IS i.array*shift:
  INT blk.mov:
  SEQ
    blk.mov := i.min*j.min              --how much to move at once
    IF
      shift<0                                       --shift one way
        SEQ
          VAL INT blocks.to.move IS
                                  ((-1)*shift.blk)/blk.mov:
          IF
            blocks.to.move>0                         --move blocks
              SEQ i=0 FOR blocks.to.move
                move(array,buffer,blk.mov,
                     (local.size-blk.mov),0,shift,
                     local.size,up.in,down.out)
            TRUE
              SKIP
          VAL INT left.over IS ((-1)*shift.blk) REM blk.mov:
          IF
            left.over>0                            --move remainder
              move(array,buffer,left.over,
                   (local.size-left.over),0,shift,
                   local.size,up.in,down.out)
            TRUE
              SKIP
      shift>0                                     --shift other way
        SEQ
          VAL INT blocks.to.move IS (shift.blk)/blk.mov:
          IF
            blocks.to.move > 0                       --move blocks
              SEQ i=0 FOR blocks.to.move
                move(array,buffer,blk.mov,0,
                     (local.size-blk.mov),shift,
                     local.size,down.in,up.out)
            TRUE
              SKIP
          VAL INT left.over IS (shift.blk) REM blk.mov:
          IF
            left.over>0                            --move remainder
              move(array,buffer,left.over,0,
                   (local.size-left.over),shift,
                   local.size,down.in,up.out)
            TRUE
              SKIP
      TRUE
        SKIP
:
```

Figure 5-26
Code for shifting a data set upward or downward

The move routine is called one final time with left.over used in place of blk.mov.

If the shift value is greater than zero, the same data blocks as before are passed upward. The only difference is that the pointers to the data to be sent and received are reversed.

Performance Issues

Using this shift routine, we can perform the same analysis as we did for the expanded data set approach. With the same nine-element kernel on a variety of array sizes, we can measure the calculation time for the convolution implemented with array shifts, and then compare this time with the time needed to perform the same convolution arithmetic but without doing any array shifts.

The results of this comparison are shown in Fig. 5-27. The overall efficiencies are much lower than in the expansion example. This is because the local data is being continually relocated in memory as well as being communicated. It is very important to realize that this data relocation is not necessary for most problems. Here the entire data set is being moved; in many applications only small subsets of the entire data set are moved.

As can be seen from the graph, the overall efficiency ranges from a high of nearly 50% to a low just above 25%. The high and low positions on this efficiency graph are reversed from the positions on the expand routine graph. As the array gets larger, the effort of moving the local data with the shift routine increases faster than the effort of doing the communication between the processors.

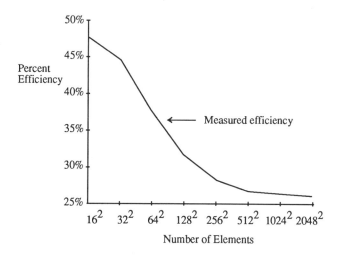

Figure 5-27
Computational efficiency for various data set sizes

In contrast to the data expansion approach, the great advantage of the shifting approach is that it uses little extra memory. Regardless of the size of the convolution kernel, the shift routine uses only enough memory to store a copy of the original data set. The data expansion routine, on the other hand, can very quickly use several times as much memory for larger kernels.

An Efficiency Comparison

It is unwise to assume that, because the array shifting approach to communicating nonlocal data is less than half as efficient as the expanding approach, the shifting method is not worth pursuing. Both examples used a nine-element kernel for the convolution; because that is a small kernel, it requires neither much extra work nor much extra storage for the expanded data set approach. Thus the expanded data approach is much more effective than the shifting approach for small kernel sizes. For larger kernels, the shifting approach can be relatively more efficient, while the expanded data set approach will require more and more memory storage as the data communication and storage overhead increases.

To illustrate this point, let us consider a set of kernels of different sizes, all of which have nine nonzero elements. Regardless of the size of the kernel, nine shifts are needed for communicating the array and computing the convolution. Thus we expect that the efficiency of the shifting approach should remain relatively constant for all of the kernel sizes. The expanded data set approach, on the other hand, will require the exchange of more and more data as the kernel size increases, and the efficiency should consequently decrease.

A demonstration of the computational efficiency of these two approaches for a range of kernel sizes is shown in Fig. 5-28. This demonstration program convolves a 1024-by-1024 array with kernels of different sizes using the shifting approach and the data expanding approach on a 10-by-12 processor toroid. The expanding approach is very efficient for small kernel sizes, but is much less efficient as the kernel size increases. The program efficiency for the expanding approach with the 257-by-257 kernel is found using a more general and faster routine than the routine listed in Fig. 5-22, since the expansion routine in Fig. 5-22 cannot expand data sets for kernel sizes larger than the amount of data stored in the immediate neighbors of each processor.

The shifting approach, in contrast to the expanded data approach, does stay relatively constant in efficiency as the kernel size changes. Efficiency actually improves somewhat as kernel sizes increase up to 257-by-257. This is because the processors find it faster to communicate blocks of data with the links than to move the data internally in memory. For the 257-by-257 kernel, data must be passed across two processors in each direction, and the efficiency decreases.

From this demonstration we see that the expanding approach is more efficient than shifting for every kernel size. Not only is the expanded data approach more efficient, it is also faster, requiring 0.74 seconds to compute versus 1.05 seconds for the shifted approach. But the speed and efficiency come at a great price.

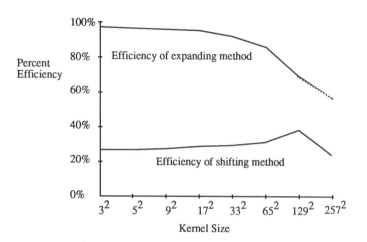

Figure 5-28
Convolution efficiency for various kernel sizes
using two approaches for a 1024 x 1024 array

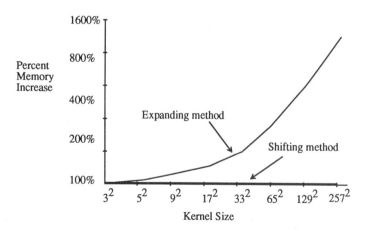

Figure 5-29
Convolution memory requirements for various kernel
sizes using two approaches for a 1024 x 1024 array

The shifted approach requires 8 MBytes of storage for two 1024-by-1024 arrays of four bytes per element. In comparison, the expanded data approach, on a 120-processor machine, needs about 62 MBytes of memory storage for the original 4-MByte array and the 58-MByte expanded data set. The percent memory increase required for each approach for each of the kernel sizes is graphed in Fig. 5-29.

In Summary

Data parallelism is the most popular and simplest technique for programming parallel computers. Geometrically parallel systems provide a simple way to decompose a problem into parallel pieces and provide automatic load balancing. Because the data are distributed among the processors, each processor will perform the same operation on its portion of data, so that only one program need be written. Calculating the distribution of the data among the processors is a fairly simple task, as is finding a subset of the data once it is distributed.

When a computation requires data which is not stored locally, it must be communicated from a neighboring processor. This communication can significantly reduce the efficiency of the computer, especially if the size of the data set is comparable to the size of the processor array.

Loading and unloading data into and out of a parallel processor also reduces the performance of the computer. By careful use of all of the available communication links and by double buffering the input and output communication stages in each processor, a programmer can reduce this overhead.

There are many ways to communicate data between processors when a computation requires such communication. One approach is expanding the local data set to store redundant information from neighboring processors. This method can require a great amount of extra storage if a lot of data are needed, but once the data are communicated any subsequent computation is very efficient. Communicating all of the necessary data from processor to processor without storing any of it permanently is a second alternative. Only communicating the data reduces the storage requirements but can involve more communication overhead. This second approach may often be the only practical alternative for large data sets which require information from many very remote locations. Generally, practical programs will use some combination or mixture of these two approaches to effectively meet the needs of a particular application.

Chapter 6

Deadlock-Free Routing

Routing programs are a fundamental tool for creating message-passing parallel computers. A general-purpose routing program provides a communication structure within a network so that every processor in the network can pass information to every other processor. Routing programs are especially useful for network architectures which are not highly structured and for programs in which communication between processors occurs in an irregular manner. A routing program provides a completely general method for an arbitrary collection of interconnected processors to communicate in a completely random way.

In both geometric and pipeline parallelism, the architecture of a network is tightly coupled to the communication methods of the parallel programs. In data parallelism, a problem's data are generally distributed in a way structured to match the network on which the program runs, thus minimizing the communication overhead in the parallel processor. In contrast, using pipeline processing reduces a network's communication overhead with the careful matching of an application's program structure to the processor network.

Using a general-purpose routing program frees the programmer from concerns both about the network structure and the distribution of data by providing a generic platform on which to implement virtually any application. However, such programming freedom comes at a high price; routing programs impose a substantial overhead on communication within a parallel computer. A user must trade off the efficiency advantages and programming limitations of a structured network of parallel processors for the programming advantages and efficiency limitations of an unstructured network of parallel processors.

The programs implemented in a parallel computer using message passing are generally similar in structure to those used in a processor farm (Chap. 3). In a processor farm, each processor works independently on a particular task allocated to it by a controller. No interprocessor communication is necessary except for messages passed to and from the controller. Indeed, a processor farm basically implements a simple message-passing scheme in which each processor only sends messages to, and receives messages from, the controller.

A more general routing program allows the processors to send messages to each other so that they can cooperate on computing tasks. With the use of a routing program, there is no need for a particular processor to act as a controller, since any processor can communicate with any other. In such a parallel system, the program control may be distributed over the processor network in the same way that the work is distributed.

One interesting characteristic of message-passing parallel systems is their unpredictability. It is generally impossible to know at any given time just which processors are doing what, and which messages are going where. Message-passing parallel systems really create an environment in which the many processors form an amorphous "sea of processors" which works on problems in unpredictable ways. This lack of structure can make it very difficult to find programming errors, especially when the errors may be processor-interdependent.

Message-passing systems are especially useful for parallel operating systems. Parallel operating systems must arbitrarily create and assign tasks to various processors. Therefore, such systems must have a guaranteed, deadlock-free method for communicating with any processor. This requirement can be met through the use of a message-passing program. A message-passing operating system also provides program portability from one processor network to another. As a program moves from computer to computer, the details of the routing algorithms within the operating system may change, but the way the application program interacts with the message-passing engine does not.

Program Issues

Computers that use a routing engine for interprocessor communication utilize a communication shell which runs in each processor concurrently with the application program (Fig. 6-1). This communication shell allows any processor in the network to communicate with any other processor. Each processor is given an address to distinguish it from every other processor. The address is often generated from the structure of the parallel processor architecture, but in an arbitrarily connected network, the address may also be arbitrary. If any processor should need to request information from, or pass information to, another processor, it simply sends a message to the communication shell together with the address of the destination processor.

The communication shell actually performs all of the interprocessor communication. It receives messages from the local application program and from other processors, and sends the messages on their way. If the network has a point-to-point connection between every processor, sending the messages on their way is very simple; the message is sent directly to the destination processor. However, much more frequently the network does not have connections between every processor, and the communication shell must first forward the message to some other processor before the message reaches its final destination. The following discussion presents programs for use on systems which are *not* totally connected.

Any program shell which implements a general message-passing scheme for interprocessor communication must ensure that a message will ultimately be delivered to its destination. Furthermore, it is likely that every processor in a network will be generating messages at the same time, so the routing engine must also en-

Chapter 6   *Deadlock-Free Routing*

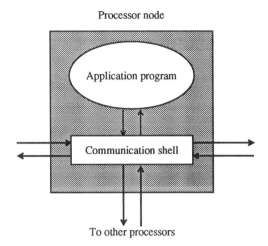

Figure 6-1
An application program and communication shell within a processor

sure that the network will not become completely clogged with messages and grind to a halt, unable to pass any messages at all.

The first requirement, guaranteeing message delivery, is fairly easy to meet. To do so, the routing algorithm must simply avoid the creation of closed loops when calculating the path which a message must take in going from one processor to another. Every step a message takes in moving from one processor to another must move it logically closer to its final destination processor.

The choice of routing algorithm, which determines the interprocessor communication path for a particular message, obviously depends on the shape of the network on which the communication shell is running. But even within a given network there are many ways a message can be routed, and many different algorithms exist which optimize the routing engine for different factors, such as distance traveled, load balancing, message latency, and so on.

A routing algorithm can be dynamic, so that the particular route a given message travels is not the same each time the message is sent. As message traffic conditions within the network change, it is possible to reroute a message to avoid heavier message traffic and thus distribute the communication load more evenly over a network, increasing the network's efficiency.

The second requirement for routing engines is more difficult to meet. Any communication and routing scheme must be able to guarantee that it will not deadlock, that is, under any conditions whatsoever the communication program must be able to continue moving messages closer to their destinations. Although as message traffic increases it is inevitable that network communications will slow down, the messages must be able to keep moving closer to their destination. It must be impossible for any group of messages to mutually halt their own progress, no matter how many messages there are, what order they are sent in, what processors they are from, or to what processors they are sent.

As an example of a deadlocked network, consider a ring of processors around which messages can pass in one direction. The communication shell within each processor is made up of two buffers, one at the input from a link and one at the output. A message is input at one buffer, and, if that message is destined for another processor, passed to the output buffer. Figure 6-2 shows a four-processor ring trying to pass a series of messages from each processor to the processor two to the right. The curved links on the ends of the ring indicate that the links are connected around the two ends of the ring. Four stages of this process are shown with four successive rings, beginning with the top ring. An empty box in a processor indicates an empty buffer; a number with a subscript in a box indicates a message stored in the buffer. The number is the destination address for the message, and the subscript labels the messages within the series. An arrow pointing into an output buffer indicates that an application program is sending a message to the communication shell.

Initially, at time 0, all of the buffers in the ring network are empty. At time 1, the application program in every processor tries simultaneously to send a message, filling all of the output buffers. At time 2, each output buffer passes its message to the right and receives a new message from the application program. At time

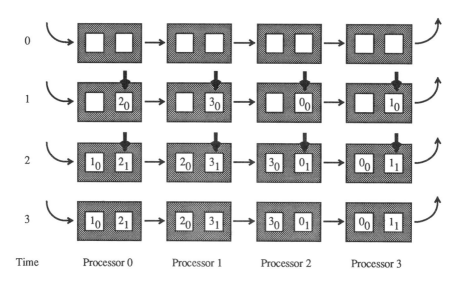

Figure 6-2
A four-processor ring becoming deadlocked

3, the processors are deadlocked because every buffer is full and therefore none is available to receive the message that another buffer is trying to pass. Because the input buffers are full, the output buffers cannot pass any messages to them. The input buffers cannot pass any messages because the output buffers are full and none of the messages are at their destination processors yet. Because the output buffers are full, the application program cannot pass any messages either. Each buffer is waiting for the following buffer to clear, and together the buffers form a closed loop, causing the network to deadlock. If the loop is broken, messages can continue on their way. It is true that by executing the application at a lower priority than the communication process, we can reduce the likelihood of deadlock. Nevertheless, executing the communication shell at a higher priority than the application process cannot guarantee that deadlock will not occur.

There are many routing algorithms which create deadlock-free message-passing programs, and there are many researchers looking for new and better schemes. The one we will implement here uses a structure of virtual links to prevent the construction of closed loops. The algorithm for this method was published by William Dally and Charles Seitz in 1987.[†]

This algorithm implements deadlock-free routing in rings by using a pair of virtual channels communicating over a single hardware link (Fig. 6-3). Each processor will pass messages on one channel or the other depending on its own address and the destination address of the message it is sending. The choice of channels in each processor is arranged so that it is impossible to form a closed loop of channels.

When routing a message, a processor chooses a channel by comparing the destination address of the message with its own address. If the destination address is greater than the processor's own address, the message is routed on the upper channel; if the destination address is lower than its own, the message is routed on the lower channel. Therefore, any message coming from processor 0 must use the upper channel and any message coming from processor 3 must use the lower chan-

---

† W. Dally and C. Seitz, "Deadlock-Free Message Routing in Multiprocessor Interconnection Networks," *IEEE Transactions on Computers*, vol. C-36, No. 5 (May 1987), pp. 547-553.

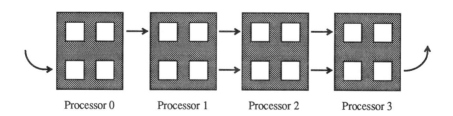

Figure 6-3
A ring of processors with two virtual channels

nel, and the closed loop created with a single communication channel is opened. Because processor 3 cannot use the upper channel and processor 0 cannot use the lower channel, Fig. 6-3 does not show them. No set of messages can be mutually interdependent on progress in one channel since the lower channel will never deliver a message to the upper channel. Messages can move only along one channel or from the upper channel to the lower one.

To demonstrate this freedom from deadlock, let us reconsider the earlier example of deadlock using the virtual channel structure. Figure 6-4 illustrates a four-processor system connected in a ring and passing messages around the ring to the right. Each processor is connected to its neighbors with two channels and has a buffer in which to read data from, and write data to, the channels. Again, every processor repeatedly attempts to send a message to the processor two to the right. Each processor must use the correct channel according to the algorithm given. The processor channels which cannot be used are shown in a lighter gray.

At the beginning of the six-stage illustration (step 0), all of the buffers are empty. In step 1, every processor sends its first message. Once again, the message destination is given by the number in the box representing the buffer in which the message is written, and the message's position in the sequence of messages by the subscript. The vertical arrow represents an application program writing a message into a buffer. Processors 0 and 1 will write their messages into the upper channel's buffer because their addresses are lower than the destination processors at 2 and 3 respectively. Processors 2 and 3 will use the lower channel's buffer because their addresses are higher than the destination processors at 0 and 1.

In step 2 of the six-stage illustration, every processor sends a second, identical message at the same time that each of the first messages moves to the next processor. No message has been blocked yet. At step 3, however, only the messages sent to processor 3 can progress. The lower channel in processor 2 is blocked, but a message for processor 3 will use the upper channel, thereby preventing deadlock. The same thing happens at steps 4 and 5; messages for processor 3 continue to progress while all other links are blocked.

Except for the messages for processor 3, all messages will continue to be blocked until processor 1 stops sending messages to processors on its right. If processor 1 then decides to send a message to processor 0, the message will travel by the lower channel, freeing the message for processor 2 to continue on its way. Thus the network will always be able to advance at least one message, regardless of destination or source, providing for a deadlock-free communication network.

One-Way Virtual Channels

In order to actually construct a deadlock-free communication network on a ring using the virtual channel approach, two channels must be multiplexed onto the one hardware link connecting each pair of processors in the ring. This must be done in such a way that if communication on one channel becomes blocked, the other channel can still proceed. The two channels must appear to be two completely dif-

Chapter 6 Deadlock-Free Routing

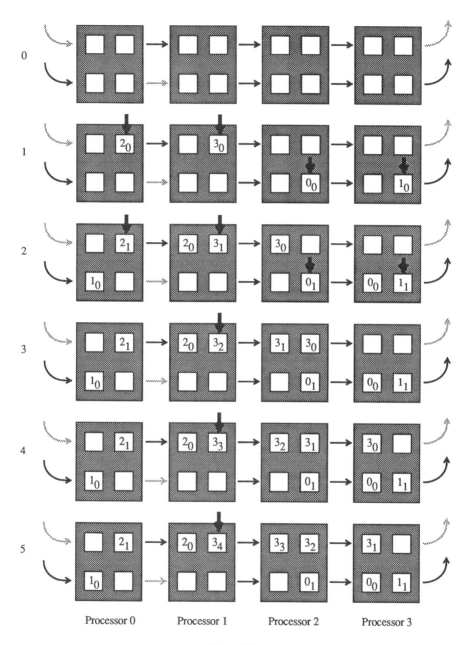

Figure 6-4
A ring of processors with two virtual channels and no deadlock

```
[2]CHAN OF MSG in,out:
PAR i=0 FOR 2
  WHILE TRUE
    SEQ
      in[i] ? message
      out[i]! message
```

Figure 6-5
Virtual channel software model

ferent devices. This virtual link structure can be supported in either hardware or software. Newer transputer designs support virtual link hardware suitable for this task, but older ones (T8xx or earlier) do not.

As an example of a software solution, the software model emulated by a virtual link multiplexing program is shown in Fig. 6-5. This code simply establishes two channels which read and write data in parallel. If one channel is blocked, the other can easily continue. Our task is to create a software shell in which these two channels can pass messages over the single, bidirectional hardware link. This shell can be extended to handle any number of virtual channels, allowing two processors to communicate with as many channels as a programmer might desire.

In order to implement the software shell, at least two communication handlers must be created, one to output messages to a link (an output handler), and one to input messages from a link (an input handler). If, after receiving a message, the input handler cannot pass the message onward, the output handler must not accept another message for the same virtual channel, but, at the same time, must be free to accept a message for the other channel. This program structure will emulate the blocking of one single channel while preserving the freedom of the other. Each processor must include both an input and an output handler for reading and sending messages.

Figure 6-6 is a diagram of one way a communication handler might be constructed. The labels on the channels in the diagram correspond to the names used in the program itself. The large gray bar down the center of the figure separates the left processor from the right processor. On the left side, an output handler accepts messages from two channels and multiplexes them across the single hardware link to the right-side processor. Once a message is passed on a channel, the output handler must not accept another until it receives an "all clear" message from the right-side processor. The right-side processor consists of four parallel parts. The first (upper left box) reads in messages and passes them to the two buffers. After the buffers successfully pass their messages along, they pass a token to the return handler (bottom box) which then sends it to the output handler in the left-side processor. Once the output handler has received the token, it knows that the message has been successfully output by the buffer, and that it is free to receive another message.

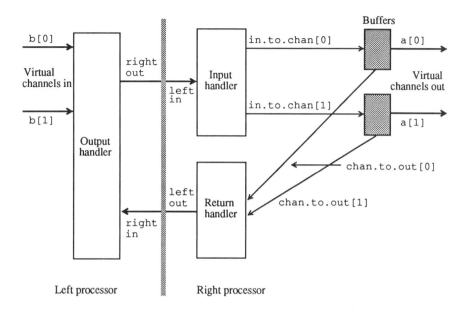

Figure 6-6
Block diagram of a one-way virtual link implementation

Each buffer process must run in parallel with the other two processes in the input handler, because if one of them fails to output a message, the other processes must remain free to pass messages on the other channel. These buffer processes are the only ones which can be blocked, while waiting for a message to output. The return handler must also execute in parallel with the input handler; otherwise it is possible that the right-side processor will be trying to return a token to the left-side processor at the same time that the left-side processor is trying to send a message to the right-side processor. If this happens, the processors will deadlock.

The code for a process which runs this communication shell is listed in Fig. 6-7. In this program, a protocol for the link communication is defined first. This protocol has four parts: first, an integer representing the virtual channel along which the message should pass; second, an integer for the destination address; third, an integer count of the message length; and finally, an array containing the message.

The process itself is called with parameters for the size of the ring, x.trans, the address of the processor, i.trans, and the input and output links to the left and right. Channels left.in and right.out are defined with the link protocol, while the right.in and left.out channels, which only communicate tokens, are defined with a simple INT protocol. After the channel arguments, the next statement in the program defines the virtual channel protocol,

```
PROTOCOL LNK IS INT;INT;INT::[]INT:
PROC shell(VAL INT x.trans,i.trans,
      CHAN OF LNK left.in,right.out,
      CHAN OF INT left.out,right.in)
  PROTOCOL MSG IS INT;INT::[]INT:
  VAL INT max.size    IS 32:
  VAL INT num.io.chans IS 2:
  [num.io.chans]CHAN OF MSG a,b:
  PAR
    ... Output handler                    Fig. 6-8
    ... Input  handler                    Fig. 6-9
    ... Buffer handler                    Fig. 6-9
    ... Return handler                    Fig. 6-9
:
```

Figure 6-7
Code for virtual channel process

which is identical to the link protocol except for the deletion of the initial channel value. The maximum size of the message is arbitrarily defined at 32, and the number of virtual channels in each direction (num.io.chans) is set at 2. Channel array a is used for the output of the virtual channels and array b is used for the input. The program then creates four parallel processes, one each for the output handler, the input handler, the buffer processes, and the return handler.

The output handler in this multiplexing system is listed in Fig. 6-8, and begins with an abbreviation of the virtual input channels to chan.in, and the definition of the two boolean variables ok. If ok is TRUE, the output handler is free to accept an input on the same virtual channel; if ok is FALSE, an input must not be attempted. The output handler then continues by initializing the ok variables to TRUE and entering an infinite WHILE TRUE loop.

This infinite loop waits for an ALT input from any one of three sources: from either of the two virtual channels, or from the hardware link. The two virtual channels are guarded by the ok variables to prevent an input on a channel which is not free. If a message from a virtual channel is received, the message, prefixed with the channel number, is passed to the output channel right.out, and the corresponding ok variable is set to FALSE. If the output handler receives an input from the right-side processor, the input must be a token and indicates that the handler is free to accept another message on the corresponding virtual channel. The output handler then sets the appropriate ok variable to TRUE.

The input, buffer, and return handlers for the multiplexer are all listed in order in Fig. 6-9. Two arrays of channels connect the processes, one array of channels using a message protocol, the other using a simple INT protocol for tokens. Included in the Fig. 6-6 block diagram are the channel names. The input handler executes a simple infinite loop, reading messages from the left-side processor on

Chapter 6   *Deadlock-Free Routing*

```
[num.io.chans]CHAN OF MSG chan.in IS
                           [b FROM 0 FOR num.io.chans]:
[num.io.chans]BOOL ok:
SEQ
  SEQ i=0 FOR num.io.chans
    ok[i]:=TRUE
  WHILE TRUE
    ALT
      INT length,x.addr:
      [max.size]INT data:
      ALT i=0 FOR num.io.chans           --channel input
        ok[i] & chan.in[i]? x.addr;length::data
          SEQ
            right.out!   i;x.addr;length::data
            ok[i]:=FALSE
      INT channel:                       --return token
      right.in? channel
        ok[channel]:=TRUE
```

Figure 6-8
Output handler for one-way virtual channels

```
[num.io.chans]CHAN OF INT chan.to.out:
[num.io.chans]CHAN OF MSG in.to.chan:
PAR
  [num.io.chans]CHAN OF MSG chan.out IS
                            [a FROM 0 FOR num.io.chans]:
  WHILE TRUE                              --input handler
    INT length,channel,x.addr:
    [max.size]INT data:
    SEQ
      left.in?    channel;x.addr;length::data
      in.to.chan[channel]! x.addr;length::data
  PAR i=0 FOR num.io.chans              --buffer handler
    INT length,x.addr:
    [max.size]INT data:
    WHILE TRUE                 --buffers, block here only
      SEQ
        in.to.chan[i] ? x.addr;length::data
        chan.out[i]   ! x.addr;length::data
        chan.to.out[i]! i
  WHILE TRUE                              --return handler
    INT channel:
    ALT i=0 FOR num.io.chans
      chan.to.out[i]? channel
        left.out! i
```

Figure 6-9
Input, buffer, and return handler code for one-way virtual channels

```
WHILE TRUE
  ALT i=0 FOR num.io.chans
    input[i]? destination;message
      output[route(destination)]! destination;message
```

Figure 6-10
An example of a poorly designed routing engine

channel `left.in`, and then passing them to the appropriate buffer, depending on the value of the virtual channel prefix sent by the output handler in the left-side processor.

Running in parallel with the input and return handlers, the buffer processes simply read messages from the input handler and attempt to output the messages on their virtual channels. When successful, they pass a token to the return handler. The return handler waits for tokens from either buffer; when a message is received, it is returned to the output handler in the left-side processor, completing a communication loop from the left-side processor, to the right, and back again.

A One-Way Ring Router

A virtual channel multiplexer may be convenient for processes which need the illusion that more than one link actually connects two processors, but it does not constitute a routing engine. In a deadlock-free router, there must be another process to read messages from the channels, decide where they should go, and send the messages there. A good routing engine, moreover, will maintain independence among the various channels in a processor, so that if one of them becomes blocked, the others can continue.

An example of a routing engine which does not exhibit channel independence is shown in Fig. 6-10. This engine must pass messages from a set of input channels to a set of output channels, and route them according to the `route` function. In this case, if the router receives a message on one channel, and is blocked trying to send the message out on another, the entire routing engine is stopped. Any other input channel will be neglected until the first output succeeds.

To avoid stopping the entire routing engine if one output is blocked, we might try constructing the engine with parallel processes, each of which is responsible for a single input. Figure 6-11 illustrates this approach. Unfortunately, this program allows two parallel processes to output a message on the same channel at the same time. Indeed, such a code fragment will not even compile without reporting an error.

In order to properly construct a legal routing engine with channel independence, we must create a parallel process for each input and each output channel. Furthermore, a practical machine must include a mechanism for communicating with an application process which takes in and sends out messages. Figure 6-12 is

Chapter 6 — Deadlock-Free Routing

```
PAR i=0 FOR num.io.chans
  WHILE TRUE
    SEQ
      input[i]? destination;message
      output[route(destination)]! destination;message
```

Figure 6-11
A second example of a poorly designed and illegal routing engine

a diagram of such a routing engine, capable of reading messages from two input channels and sending them to either of two output channels. This router uses five parallel processes, one each for the two input buffers, one each for the two output buffers, and one for the application process. The input processes are connected by channels to both output processes and to the application process. In addition, the application process can pass messages to either of the two output processes. If any one of the process communications is blocked, the other free processes can still continue to pass messages from one process to another.

Upon input of a message, the routing engine must decide which output process the message should be routed to. The application engine must make the same decision when sending a message out.

The code implementing the input and output buffers for a simple routing engine is listed in Fig. 6-13. This code creates two groups of parallel processes, one each for the input and the output. All of the processes are connected by a two-dimensional array of channels called r. The first coordinate of the array determines the input channel, and the second coordinate determines the output channel. Since

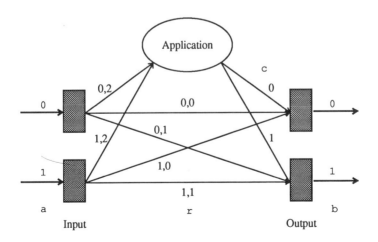

Figure 6-12
A one-way routing engine

Deadlock-Free Routing   Chapter 6

```
        PAR i=0 FOR num.io.chans                --input to router
          INT length,x.addr:
          [max.size]INT data:
          WHILE TRUE
            SEQ
              a[i]?              x.addr;length::data
              r[i][route[x.addr]]! x.addr;length::data
        PAR i=0 FOR num.io.chans                --output to channel
          INT length,x.addr:
          [max.size]INT data:
          WHILE TRUE
            ALT
              ALT j=0 FOR num.io.chans          --from router
                r[j][i]?  x.addr;length::data
                  b[i]!   x.addr;length::data
              c[i]?       x.addr;length::data   --from application
                b[i]!     x.addr;length::data
```

Figure 6-13
A one-dimensional routing engine

the application program is also considered to be a message destination, three output channels are actually needed; the channel array connecting the application process with the output buffers is called c. Each channel name and subscript is marked on Fig. 6-12. The channels for the input and output buffers are abbreviated from the same a and b channels used in the virtual channel program.

The input processes simply read messages from the outside world and pass them to the appropriate destinations, determined by the value of route[x.addr]. It is the route array which actually implements the routing function necessary for sending messages to the correct destination. The array returns the value zero, one, or two depending on whether the messages must go to output buffer zero, output buffer one, or the application process. Meanwhile, the output processes wait in an ALT construct for messages from either of the input processes or the application process. When a message is received, it is output. An application process which only reads in messages is shown in Fig. 6-14. This process reads messages from any of the input buffers, but does not respond to them. Any useful system, of course, would have to supplement the input code with an application task.

For a network whose processors do not include virtual link hardware, the routing engine just described can be connected to the virtual channel multiplexer discussed earlier, creating a complete router. Figure 6-15 shows the complete top-level code necessary for doing this; the listing is a superset of the code shown in Fig. 6-7. Figure 6-15 also includes c channels for communicating between the application process and the output buffers, and an array of r channels for the router. Variable x.trans defines an array of variables called route, which holds the routing information for every destination.

```
    WHILE TRUE
      INT length,x.addr:
      [max.size]INT data:
      ALT i=0 FOR num.io.chans           --read everything
        r[i][num.io.chans]? x.addr;length::data
          SKIP                           --do nothing
```

Figure 6-14
An application process which simply accepts messages

```
PROTOCOL LNK IS INT;INT;INT::[]INT:
VAL INT x.trans IS 13:
PROC work(VAL INT i.trans,
      CHAN OF LNK left.in,right.out,
      CHAN OF INT left.out,right.in)
  PROTOCOL MSG IS INT;INT::[]INT:
  VAL INT max.size    IS 32:
  VAL INT num.io.chans IS 2:
  [num.io.chans]CHAN OF MSG a,b,c:
  [num.io.chans][num.io.chans+1]CHAN OF MSG r:
  [x.trans]INT route:
  PROC router([]INT route)              --route algorithm
    SEQ i=0 FOR x.trans
      IF
        i = i.trans                     --me
          route[i]:=num.io.chans
        i > i.trans                     --on right
          route[i]:=1
        TRUE                            --on left
          route[i]:=0
  :
  SEQ
    router(route)
    PAR
      ... Application handler           Fig. 6-14
      ... Output      handler           Fig. 6-8
      ... Input       handler           Fig. 6-9
      ... Buffer      handler           Fig. 6-9
      ... Return      handler           Fig. 6-9
      ... Route       handler           Fig. 6-13
:
```

Figure 6-15
Complete code for a one-way, deadlock-free routing engine

At the beginning of the routing engine program, the `route` table is initialized in process `router`. This process uses the algorithm discussed earlier in which messages with a destination address greater than the local processor's address are routed on output channel 1, and messages with a destination address smaller than the local processor's address are routed on output channel 0. If the destination address is equal to the local processor's address, the message has arrived and is sent out on channel 2, which is connected to the application processor.

Once `route` is initialized, the program begins running in earnest. The application process, the output, input, buffer, and return handlers, and the routing engine are all run in parallel. The application process creates and absorbs messages, interacts with the routing engine, and ignores the virtual link processes. It is important to remember that the application process must guarantee that any message sent to it is eventually consumed. Therefore, the application process must not become blocked when sending a message to an output buffer of the router.

As an example of what can happen if the application handler is not free to read in messages, consider a hypothetical and pathological relationship between two processors in which each starts by sending a message to the other. For every message received, each processor replies with two more messages. Eventually, all of the communication buffers between the two processors will be filled up. Unless both processors can continue to receive messages despite their inability to send messages, the two processors will deadlock. This problem can be avoided if the input from the network is done in parallel with the output to the network, and if adequate message buffering is provided.

Two-Way Virtual Channels

The one-way virtual channel multiplexer described in the previous section can be expanded for use in a bidirectional system. Using a bidirectional routing engine will obviously shorten the distance any message must go, since the message can travel in either direction around the ring. On the ring example we have been considering, the average interprocessor distance in the ring is halved if messages can travel in either direction.

A bidirectional channel multiplexer must incorporate both the input and output processes used in the one-way example. Each processor will need one complete multiplexer for each link used. Figure 6-16 is a diagram of the two-way channel multiplexer. The multiplexer includes an input handler, an output handler, two output buffers, and a delay buffer. Message protocol channels connect the input handler to the output buffers. Token protocol channels connect the output buffers to the output handler, the input handler to the delay buffer, and the delay buffer to the output handler.

The input handler in the bidirectional channel multiplexer waits for data from the neighboring processor. If it receives a message, the message is passed to the output buffers; if it receives a token, the token is passed through the delay buffer to the output handler. The output handler monitors the virtual channels, the output

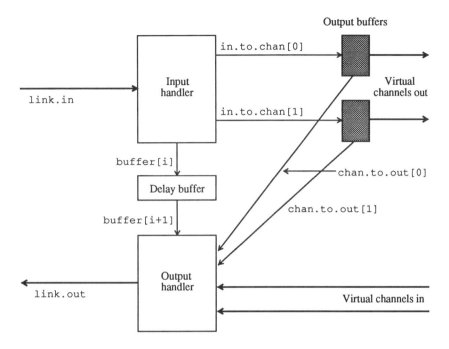

Figure 6-16
A bidirectional virtual channel handler

buffer token channels, and the delay buffer channel. When a message arrives from a virtual channel, it is passed to the neighboring processor. The link on which the message arrived is then shut down until the neighboring processor replies with a token indicating that the message has been successfully sent on its way. Thus, data passes from the output handler to the neighboring processor's input handler, to the neighboring processor's output buffer, which returns a token to its output handler. The neighboring processor's output handler returns the token to the local input handler. The token then passes through the delay buffer and back to the original output handler. This entire cycle is fundamentally similar to the one-way example, the only real difference being that any link communication is one of two types, either a returned token or a message. The handlers must be able to distinguish the two communication types appropriately.

The delay buffer must be included between the input and output handlers so that a new potential deadlock problem will be prevented. Assume that the processors on both sides of a link have sent each other a message, and that they are about to do so again. At this point each processor will be trying to send a second message to the other as well as trying to return a token in response to the first message. It is possible that the output handlers of both processors will be attempting an output on

```
PROC two.way(CHAN OF LNK link.in,link.out,
            []CHAN OF MSG chan.in,chan.out)
  [num.io.chans]CHAN OF INT chan.to.out,buffer:
  [num.io.chans]CHAN OF MSG in.to.chan:
  PAR
    [num.io.chans]BOOL ok:                         --output handler
    SEQ
      SEQ i=0 FOR num.io.chans
        ok[i]:=TRUE
      WHILE TRUE
        ALT
          INT length,x.addr:
          [max.size]INT data:
          ALT i=0 FOR num.io.chans              --input from channel
            ok[i] & chan.in[i]? x.addr;length::data
              SEQ
                link.out! msg; i;x.addr;length::data
                ok[i]:=FALSE
          INT channel:
          ALT i=0 FOR num.io.chans                 --input token
            chan.to.out[i]? channel
              link.out! ret; i
          INT channel:                          --input from delay
            buffer[num.io.chans-1]? channel
              ok[channel]:=TRUE
    INT length,channel,x.addr:                    -input handler
    [max.size]INT data:
    WHILE TRUE
      link.in? CASE
        ret; channel
          buffer[0]! channel
        msg; channel;x.addr;length::data
          in.to.chan[channel]! x.addr;length::data
    INT channel:                                  --delay buffer
    PAR i=0 FOR num.io.chans-1
      WHILE TRUE
        SEQ
          buffer[i]?    channel
          buffer[i+1]!  channel
    PAR i=0 FOR num.io.chans                    --output buffers
      INT length,x.addr:
      [max.size]INT data:
      WHILE TRUE
        SEQ
          in.to.chan[i] ? x.addr;length::data
          chan.out[i]   ! x.addr;length::data    --can block here
          chan.to.out[i]! i
:
```

Figure 6-17
Listing for a bidirectional virtual channel handler

the link at the same time that their respective input handlers are trying to pass a token to them. If the delay buffer were not placed between the input and output handlers, this situation would result in deadlock. If more virtual channels are added to the system, the buffer must grow, so a simple first-in, first-out (FIFO) buffer of size num.io.chans−1 is used to delay the tokens being returned. Using a single long buffer rather than multiple parallel buffers forces the system to handle the tokens in the order in which they are received.

A process implementing the bidirectional channel multiplexer is listed in Fig. 6-17. The process is called with the hardware input and output links, as well as an array of virtual input and output channels. Note that the link protocol (listed at the beginning of Fig. 6-18) is more complex than the simple protocol used up to this point. It now has two cases, the first for messages and the second for tokens. After the PROC statement is made, the local channels connecting the input handler to the output buffers, and the output buffers to the output handler, are defined. The process begins execution with a PAR construct running four parallel processes: the output handler, the input handler, the delay buffer, and the output buffers, listed in that order.

The output process for the bidirectional channel multiplexer is slightly different than the output handler discussed earlier. The process no longer waits for a token from the hardware link, but instead waits either for a token to be returned from an output buffer or for a token to be passed from the delay buffer. The new output handler must also insert the appropriate link protocol token for both the message output and the token return output.

The input process for the bidirectional multiplexer also differs from the earlier example in that it must be prepared to receive either messages or tokens. The two possibilities are distinguished by a CASE statement. As before, messages are forwarded to the output buffers, but tokens are now passed to the delay buffer. The delay buffer simply reads in the token value and passes it on to the output handler, which will then use the token to free up a virtual input channel. The output buffer processes are just like the ones used in the one-way example.

A Two-Way Ring Router

Using the bidirectional virtual channel handler, constructing a routing engine which can pass data from a source processor to any destination processor in either direction around a ring is quite a straightforward task. Figure 6-18 lists the top-level code for creating such a router. This routine is very similar to the one in Fig. 6-15 (the original, one-way communication shell), with just a few minor differences. The new value num.rt.chans is double num.io.chans and reflects the fact that there are now four input and output channels for the router to deal with (two in each direction), while the channel arrays depending on these constants automatically scale in size appropriately. The r array connects all of the input buffers to all of the output buffers and the application process, just as it did in the one-way router case.

```
PROTOCOL LNK
  CASE
    msg;   INT;INT;INT::[]INT
    ret;   INT
:
VAL INT x.trans IS 13:
PROC work(VAL INT i.trans,
      CHAN OF LNK left.in,right.out,left.out,right.in)
  PROTOCOL MSG IS INT;INT::[]INT:
  VAL INT max.size     IS 32:
  VAL INT num.io.chans IS 2:
  VAL INT num.rt.chans IS num.io.chans*2:
  [num.rt.chans]CHAN OF MSG a,b,c:
  [num.rt.chans][num.rt.chans+1]CHAN OF MSG r:
  [x.trans]INT route:
  ... PROC router                                Fig. 6-19
  ... PROC two.way                               Fig. 6-17
  SEQ
    router(route)
    PAR
      ... Application handler                    Fig. 6-14
      two.way(right.in,right.out,
                       [b FROM num.io.chans FOR num.io.chans],
                       [a FROM num.io.chans FOR num.io.chans])
      two.way(left.in,left.out,  [b FROM 0 FOR num.io.chans],
                                 [a FROM 0 FOR num.io.chans])
      ... Route handler                          Fig. 6-13
:
```

Figure 6-18
Listing for a bidirectional virtual channel routing engine

```
  PROC router([]INT route)
    VAL INT half IS x.trans/2:
    INT direction:
    SEQ i=0 FOR x.trans
      SEQ
        IF
          (i > (i.trans+half)) OR              --wrapped around
              ((i < i.trans) AND (i >= (i.trans-half)))
            direction:=0
          TRUE
            direction:=2
        IF
          i = i.trans                                    --me
            route[i]:=4
          i > i.trans                                    --right
            route[i]:=1+direction
          TRUE                                           --left
            route[i]:=0+direction
  :
```

Figure 6-19
Listing for a bidirectional routing algorithm

The main process in the two-way ring router creates four parallel processes: the application handler, the routing engine, and two bidirectional channel multiplexers. The application and routing engines are identical to the ones listed earlier, although the number of router processes in this case scales with the number of input and output channels. Mutually exclusive subsets of the input and output channels are passed to the two channel multiplexers. The first handler receives the left-side links and the virtual channels subscripted from zero to one; the second handler receives the right-side links and the virtual channels subscripted from two to three. These virtual channels must then be correctly matched to the links connected by the routing algorithm.

Although the routing engine code in the two-way case does not change from the one-way case, the routing algorithm (shown in Fig. 6-19) is more complicated. For a complete routing table to be created, each processor in the network must choose a path on which to send a message to any other processor. The variable `half` is defined to be one-half the ring size and is used to calculate the direction in which to pass messages. The first IF statement in the router process makes this decision. The logical test determines whether the destination processor is more than halfway around the ring to the right or less than halfway around the ring to the left. Both tests must be made because of the discontinuous processor numbering between processor 0 and its left-hand neighbor. If the destination is on the left, the message should be passed to the left and `direction` is set to 0; if the destination processor is closer on the right, `direction` is set to 2. The `direction` value *must* match the pair of channels passed to the handlers in the main PAR statement. In Fig. 6-19, channels 0 and 1 are used to pass messages to the left and channels 2 and 3 to pass messages to the right.

Once the direction to a processor is determined, one of the two virtual channels going in that direction must be chosen. As before, we use the high channel if the destination processor has a higher address than the local processor, and the low channel otherwise. The same criterion is used for messages passed in either direction, since the actual choice of high or low channel for passing messages to either the left or the right is irrelevant as long as every processor chooses consistently.

A Four-Way Toroidal Router

The two-way router for a ring can be readily extended to become a four-way router for a toroid. Obviously, the major differences will be the fact that links must be connected in four directions and that the routing algorithm must take into account the two-dimensional interconnection structure. The link multiplexing code and the routing engine for the four-way router are identical with those of the two-way ring router.

Figure 6-20 is a block diagram of the resulting system running in each processor. Four handlers pass messages in the four directions: left, right, up, and down. Each handler multiplexes a pair of virtual channels over its link. An application routine connects to the routing engine; the routing engine directs the mes-

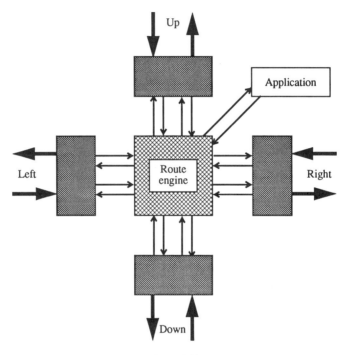

Figure 6-20
A four-way, toroidal routing shell

Chapter 6 Deadlock-Free Routing

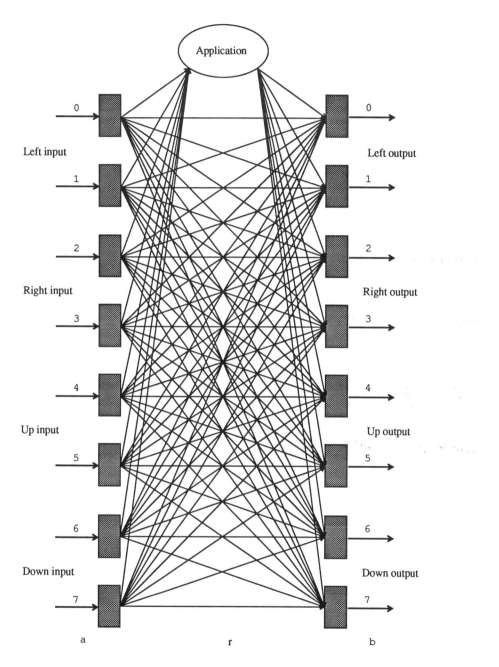

Figure 6-21
The eight-channel route engine

sage traffic to and from the handlers. So that the independence of the various interhandler channels is maintained, a channel from every input to every output is needed. Figure 6-21 is a diagram showing the necessary connections between the handlers and the application process.

The routing engine for the four-way router must connect eight input channels, eight output channels, and the application process. Every channel input must be connected to every channel output. The input channels to the router make up an array of channels called a, and the output channels make up an array called b. The handlers, of course, use a as the output and b as the input. The channel array from the application process is called c, and the channel array internal to the route engine is a two-dimensional array called r, which connects all of the inputs to all of the outputs. Notice that this structure permits a message travelling in one direction to reverse its course and head back out the way it came in. This capability may be useful for some adaptive routing schemes, but is unnecessary for the deterministic routing algorithm used here.

A program implementing this four-way routing engine is listed in Fig. 6-22. The program is, once again, very similar to the two-way example of a routing engine on a ring, but with one major difference. Each processor in the toroidal processor array uses two values, an x.addr value and a y.addr value, to describe its position in the array. These two values then become the destination address for each message, rather than the one value used in the ring. To accommodate both address values, the MSG and LNK protocols incorporate an extra INT value. Throughout the program, then, every time the x.addr parameter is used, it must be accompanied by a y.addr parameter. Since this is really a trivial programming change, the routines are not reprinted here. In addition to the use of y.addr, the num.rt.chans value must be set to eight to accommodate the eight input and eight output virtual channels in each processor.

The routing engine for the four-way toroidal router must be expanded to reflect both the new two-dimensional structure of the processor array and the increased number of channel inputs and outputs. The routine implementing the two-dimensional routing algorithm is listed in Fig. 6-23. This algorithm is similar to the one used for the ring router, but creates four paths at once, rather than two. Any message is passed vertically first until it arrives at its correct row. The message is then passed along the row until it reaches its destination processor. To implement this routing procedure, we must construct the array of routing information in two dimensions, one each for the x and y portions of the destination address. Initially, the same y routing information is placed in every x position, precluding any message routing in the x direction. This y calculation is repeated in a SEQ loop for x.trans times.

Once the y addresses are calculated, the route array can be corrected so that messages can be passed along the local row as well. For every destination processor in the local processor's row only, the route is calculated. This value is then

```
PROTOCOL LNK
  CASE
    msg;   INT;INT;INT;INT::[]INT
    ret;   INT
:
VAL INT x.trans IS 13:
VAL INT y.trans IS 10:
PROC work(VAL INT j.trans,i.trans,a,b,
      CHAN OF LNK up.in,up.out,down.in,down.out,
      CHAN OF LNK left.in,left.out,right.in,right.out)
  PROTOCOL MSG IS INT;INT;INT::[]INT:
  VAL INT max.size IS     32:
  VAL INT num.io.chans IS 2:
  VAL INT num.rt.chans IS num.io.chans*4:
  [num.rt.chans]CHAN OF MSG a,b,c:
  [num.rt.chans][num.rt.chans+1]CHAN OF MSG r:
  [x.trans][y.trans]INT route:
  ... PROC router                              Fig. 6-24
  ... PROC two.way                             Fig. 6-17
  SEQ
    router(route)
    PAR
      ... Application handler                  Fig. 6-14
      two.way(left.in,left.out,  [b FROM 0 FOR num.io.chans],
                                 [a FROM 0 FOR num.io.chans])
      two.way(right.in,right.out,
                     [b FROM num.io.chans FOR num.io.chans],
                     [a FROM num.io.chans FOR num.io.chans])
      two.way(up.in,up.out,
                     [b FROM num.io.chans*2 FOR num.io.chans],
                     [a FROM num.io.chans*2 FOR num.io.chans])
      two.way(down.in,down.out,
                     [b FROM num.io.chans*3 FOR num.io.chans],
                     [a FROM num.io.chans*3 FOR num.io.chans])
      ... Route handler  with .io set to .rt    Fig. 6-13
:
```

Figure 6-22
Code for the eight-channel route engine

```
PROC router([][]INT route)
  VAL INT y.half IS y.trans/2:
  VAL INT x.half IS x.trans/2:
  INT direction,channel:
  SEQ
    SEQ j=0 FOR y.trans                              --set up y
      SEQ
        IF
          (j > (j.trans+y.half)) OR
          ((j < j.trans) AND (j >= (j.trans-y.half)))
            direction:=4
          TRUE
            direction:=6
        IF
          j > j.trans                                --below
            channel:=1+direction
          TRUE                                       --above
            channel:=0+direction
        SEQ i=0 FOR x.trans                          --copy to all x
          route[i][j]:=channel

    SEQ i=0 FOR x.trans                              --set up x
      SEQ
        IF
          (i > (i.trans+x.half)) OR
          ((i < i.trans) AND (i >= (i.trans-x.half)))
            direction:=0
          TRUE
            direction:=2
        IF
          i = i.trans                                --me
            channel:=num.io.chans
          i > i.trans                                --right
            channel:=1+direction
          TRUE                                       --left
            channel:=0+direction
        route[i][j.trans]:=channel
:
```

Figure 6-23
Route algorithm for the eight-channel route engine

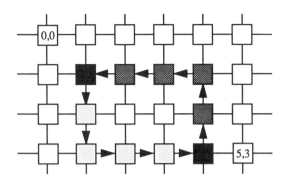

Figure 6-24
Path taken by messages moving between processors (4,3) and (1,1)

placed into the routing array with the y address of the local processor (j.trans) used to subscript the y dimension of the route array.

This two-step calculation will ensure that any message will be passed along the column from which it originated until it reaches its destination row. The router will then pass the message along the row to its final destination processor. Figure 6-24 illustrates this two-step procedure on a six processor-by-four processor portion of a larger array of processors. Processor (4,3) will send a message to processor (1,1) by passing the message upward first, to row 1, and then horizontally to the left till it reaches its destination processor. This path is indicated by the dark shading filling the processors in the path. Processor (1,1), on the other hand, will pass a message to processor (4,3) by first sending its message down, to row 3, and then horizontally to the right. The processors on this second path are filled with a lighter shading.

```
[max.size]INT data:
SEQ k=0 FOR 1000000
  SEQ i=0 FOR x.trans
    SEQ j=0 FOR y.trans            --send to all x & y
      IF
        ((i = i.trans) AND (j = j.trans))   --except me
          SKIP
        TRUE
          c[route[i][j]]! i;j;32::data
```

Figure 6-25
Test routine for the four-way routing engine

Performance Comparisons

Each of these three routing engines has been carefully tested to establish its correctness and performance. For each routing technique, every processor was programmed to pass messages of 32 words to every other processor. The application code which accomplishes this for the four-way router is shown in Fig. 6-25. The inner loop using j is omitted for the one- and two-way route engines. This routine was run in parallel with the routine in Fig. 6-14, which simply takes in each message at its destination.

Since, in this test, messages are passed from every processor to every other processor, any regular network should have a relatively even distribution of messages moving through it at any given time. This even distribution should maximize the number of messages the network can pass at a time. If the distribution of messages is not even, however, heavier message traffic in some processors might reduce the overall communication performance of the network.

The results of a test for each of the three routing engines are shown in Fig. 6-26. In the first test, performed on a ring of 13 processors, 624 million messages were passed through the ring in one direction. The test required 29.629 hours, and 5850 messages per second were communicated, with the average message travelling six processors, or halfway around the ring. The maximum distance any message must travel in a 13-processor ring is through 12 processors, all of the way around the ring.

We might expect that the two-way ring router would perform two, or perhaps four, times faster than the one-way ring router, since messages can be sent on two links at a time instead of one, and since the distance between any two processors on a ring is halfway around the ring instead of all of the way around the ring. However, there is still only one processor executing the router program, and the more complex the program becomes, the more slowly it will execute. The one-way router requires 10 parallel processes while the two-way needs 19. As more input and output handlers are added, the ALT tests become larger as well. Thus we might hope that the two-way router will pass 2.1 times more messages than the one-way

	Number of processors	Number of messages	Total time	Messages per second
One-way ring router	13	624 million	29.629 hours	5850
Two-way ring router	13	1248 million	35.349 hours	9807
Four-way toroid router	120	7200 million	34.127 hours	58604

Figure 6-26
Route engine performance for three examples

router ($2 \times 2 \times 10/19$). However, we must remember that the processor links are only 33% faster when passing data in both directions simultaneously than when passing data in only one direction. Thus a more realistic estimate of the overall increase in performance is $2 \times 1.33 \times 10/19$, or 1.4.

The test result for the two-way router on a 13-processor ring does show that the overall performance is better than that of the one-way router. The average message in the test traveled three processors with a maximum distance of six processors, and a total of 1248 million messages were sent in 35.35 hours at an average rate of 9807 messages routed per second. The message rate is 1.7 times the rate of the one-way router, somewhat less than the optimistic estimate of 2.1 but better than the pessimistic estimate of 1.4. This result implies that the system performance for the two-way ring router is not bound by the link bandwidth, but rather by the processor node's ability to switch data.

The last network test ran with a four-way router on a 120-node toroid. A toroid has more paths available within its network than the ring, since every row and column is another ring. Thus, the overall bandwidth on a 10 processor-by-12 processor network increases by a factor of 22 over a single ring. In this example, however, not only does the routing engine become more complex, and thus slower, but the messages must also travel farther. The routing engine for the four-way router uses 37 parallel processes, nearly twice as many as the two-way ring router. The average interprocessor message distance is also greater, up to five or six processors, while the worst case distance is 11. Thus for this router we would expect that the performance will be 6.2 times faster than that of the two-way router [$(19/37) \times 22 \times (6/11)$].

The test result for the four-way router shows that the router's performance actually scales fairly well. With a test of 7200 million messages passed in 34 hours, the four-way router passes 58604 messages per second, six times as many messages as the two-way router. This result is very close to the improvement factor of 6.2 that we estimated.

It is interesting to note that this four-way message-routing scheme is not limited by the link bandwidth, but rather by the ability of the processors to switch messages. For the four-way routing test, 58604 messages of 128 bytes each are passed an average of six processors in one second, or 45 MBytes/sec. Since this is more than a factor of ten smaller than the total bandwidth available in the processor network, we can see that there is plenty of room for improvement in our deadlock-free router.

In Summary

Message-passing communication programs are a very useful tool for the implementation of parallel processing. Such programs allow a user to concentrate on the behavior of and application for each processor without having to be concerned about the complex mechanics of interprocessor communication.

Communication shells are used to implement message-passing systems and to hide the actual operation of the interprocessor communication. These shells must provide a deadlock-free means for message passing which cannot fail regardless of the volume of message traffic. In addition, good communication shells use routing engines which continue to pass messages in one direction when messages travelling in other directions may be halted.

We can create a deadlock-free router by multiplexing messages over links using virtual channels to open any possible closed communication dependencies. If the router is capable of independently passing data from any input buffer to any output buffer, the channels will not be dependent on each other, allowing messages to pass in one direction even if they cannot pass in another. While these techniques are useful in designing deadlock-free routers, they do create considerable system overhead. Nonetheless, an implementation and test of a deadlock-free router for three different networks shows that deadlock-free routers are practical and reliable for thousands of millions of messages.

Chapter 7

Worms

Of all the kinds of programs which can be executed on a network of processors, worms are one of the most interesting. Propagating themselves from processor to processor, worm programs can run on networks of arbitrary size and interconnection. Worms are especially useful for exploring and testing unknown networks of processors.

A worm program is one which begins running in one processor of a reset network and then attempts to copy itself into any connected processor. Each of the successfully copied second-generation worm programs then attempts to do the same thing to its connected processors, and the procedure goes on as the newly copied worm programs continue propagating. In the end, a network of processors is running the same worm routine in each of its component members.

Because a worm program copies itself into a network connected in any manner (that is, an arbitrarily connected network, or simply an arbitrary network), it can be used to initialize programs which run on such networks (for example, processor farms). For programs which require a regularly structured network of processors (such as a hypercube), a worm can find the size of the network structure and pass that information to the program which follows it. This capability allows the subsequent program to know the size of the network on which it is to run without that information being explicitly included in the program. Thus, this capability allows a user to write programs which can run on a network of any size, provided that the network structure is consistent with the program requirements.

Not only are worm programs useful for loading subsequent programs into networks, but they are also very useful in themselves as testers or debuggers of multiprocessor systems. In order to load a network, a worm must explore it anyway, and this exploration can serve to test or debug the network. In this case, a host processor must be connected to the network, must boot the first processor in the network, and must control the replicating worm program.

Worm programs are especially easy to construct for transputer systems because the transputer links are simple to use and because the transputer itself can be booted from data passed down any link. Thus a worm program in one processor can easily initialize another processor connected to one of its links. Indeed, the program loader in the development system used to boot a network of transputers generally boots each transputer in this way. These programs assume the use of the data links in each transputer for both communication and booting, and do not consider the use of any special monitor links which might be available on a given transputer.

Searching Strategies

Whether a worm program is used simply to explore a network or as a preliminary to another program, the worm must replicate itself throughout the network. In order to replicate itself, the worm must discover which processors connected to itself are in a reset state, boot them, and copy itself into the processors. To do this, the worm must have a strategy for deciding which processors to boot, and in what order to boot them.

To understand the different strategies, one can think of a worm program running on a network of processors as making up a tree of processors. When one processor boots another it becomes a parent and the newly booted processor is considered to be its child. The link over which the child was booted is the communication channel between the parent processor and the child processor. Any link between processors other than those between a parent and child processor are ignored and not considered part of the tree. Figure 7-1a shows a very simple binary tree, an example of a regularly structured network on which worm programs can be used. In this example, node a is the parent of nodes b and d, and nodes g and f are the children of node b. Worm programs, however, are not limited to regular networks, but can readily be used on any network, regular or irregular.

There are two basic strategies a worm program can follow when searching a tree: a depth-first search or a breadth-first search. A depth-first search simply requires that a processor not boot a second child until all of the descendants of the first child are booted. A breadth-first search differs in that a processor must boot all of its children before any of its grandchildren are booted. With nodes numbered in the order in which they are booted, Fig. 7-1b illustrates the results of a depth-first search of the tree in Fig. 7-1a while Fig. 7-1c shows the results of a breadth-first search. Both searches begin at the top of the network and examine links from left to right.

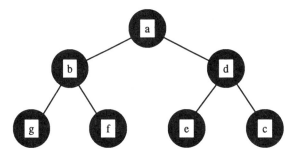

Figure 7-1a
A binary three-layer tree

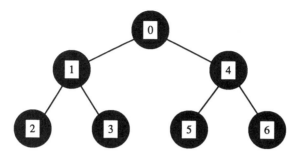

Figure 7-1b
A binary three-layer tree searched depth-first

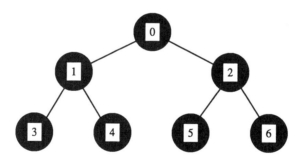

Figure 7-1c
A binary three-layer tree searched breadth-first

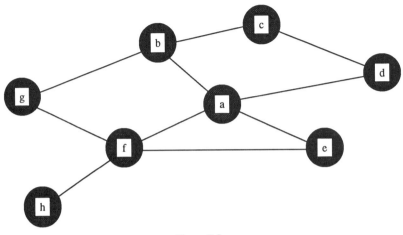

Figure 7-2a
An arbitrary network

Figure 7-2a shows an arbitrary network with arbitrarily named nodes. The order in which each node tests its connections will help determine both the final shape of the network and the order in which the network nodes are numbered. If we begin a depth-first search of this tree starting at node a and examine nodes clockwise from the left, the structure shown in Fig. 7-2b will result. The diagram is redrawn to illustrate the search order. Parent processors are drawn above child processors and connected by lines to their children. Links which are not needed for the search are not drawn. If, on the other hand, we do a breadth-first search of this tree, the structure shown in Fig. 7-2c will result.

In the binary tree shown in Fig. 7-1a, there are no extra links in the network which are not in the tree, so the order in which the links are tested is irrelevant. This, however, is not the case in Fig. 7-2a. Node a in Fig. 7-2a could choose to search node b, node d, node e, or node f first. An alternative approach (shown in Fig. 7-2d) begins at node f rather than at node a and proceeds counterclockwise from the right using a depth-first strategy.

There is a middle ground between depth-first and breadth-first searching. A particular tree may be divided into subtrees, which are portions of the larger tree. The subtrees as a group could be tested in breadth-first order while the individual subtrees themselves are searched internally depth-first. Alternatively, the subtrees could be tested in depth-first order, and each subtree examined breadth-first.

Examination of the search results in Figs. 7-2b, c, and d shows clearly that the order and search strategy used to explore a network can have a radical effect on the tree of processors constructed. Even regular networks can develop widely varying structures, as can be seen in one of the most common architectures used in parallel computers, the toroid. Figure 7-3a shows a toroidally connected processor

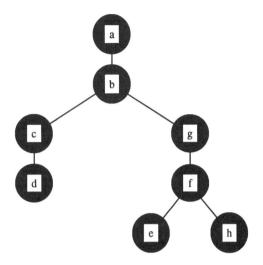

Figure 7-2b
An arbitrary network searched depth-first from node a

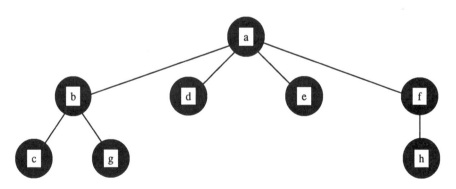

Figure 7-2c
An arbitrary network searched breadth-first from node a

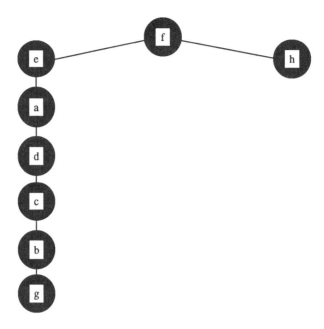

Figure 7-2d
An arbitrary network searched depth-first from node f

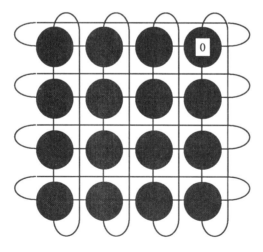

Figure 7-3a
A toroidally connected network

network. Figure 7-3b demonstrates the results of a depth-first search strategy applied to a toroid starting from the upper-right node (zero). Figure 7-3c shows the results of a breadth-first search. Nodes are labelled in the order in which they are encountered, and their links are tested from the left in counterclockwise order. The structures resulting from the two types of searches are dramatically different.

The depth-first search applied to a toroid creates a long string of nodes, each having one parent and one child. If node 0 wished to communicate with node 15 through the tree, the message would have to travel through every processor in the network. A breadth-first search produces a much shorter and wider tree. The distance from node 0 to node 15 in this case is only four processors.

This example clearly demonstrates that the breadth-first search strategy produces shorter trees and should require less communication overhead for subsequent programs which talk to the root node. However, a breadth-first search strategy is also more difficult to control and slower to run, especially for simple worms. The advantage of a depth-first search is its simplicity, in that the order of search moves from one node to a connected node, either from a parent to a child or vice versa. When a node's children all signal that they have finished exploring, the node will either attempt to boot a new child or tell its parent that it has finished exploring. At this point the parent proceeds with the search by testing its next child.

In contrast, if a breadth-first search is proceeding and a node has finished exploring, the next node to begin exploring may be far away, and almost certainly will not be a parent or a child of the node which has just finished exploring. The finished node then has no way to directly contact the next node which is supposed to begin working. The only recourse the finished node has is to either broadcast a message to all of the nodes or to pass a message up the tree of processors until a node which can properly forward the message is found. In the first case a great deal of message traffic results; in the second case some message-routing mechanism must be supported. In addition to these complications, the nodes must name or order themselves so they can pass messages to the correct processor. For the depth-first searching approach, naming or ordering is not necessary.

There is one more strategic issue which must be considered in designing a worm program. Until now, this discussion has implicitly assumed that a search program proceeds one node at a time, testing one processor after another in a strictly defined order. This is essentially a sequential search with only one processor actively testing another at a time. But each booted processor actually executes an independent program and is capable of independent action at the same time as all of the other booted processors. Thus, many of these processors could be doing tests at the same time. Such a parallel exploration should be much faster. Unfortunately it is also more complicated, since the worm will have to deal with the likelihood that two processors will try to test another processor simultaneously. Another complication of parallel exploration is that the nodes cannot be labelled in chronological order since each active processor has no idea what other processors may be do-

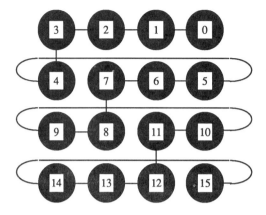

Figure 7-3b
A toroidally connected network searched depth-first

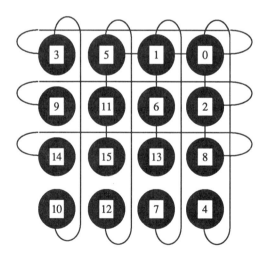

Figure 7-3c
A toroidally connected network searched breadth-first

ing and when they may be doing it. Indeed, it is possible that the same network will develop as a different tree each time it is explored.

In the remainder of this chapter, two topics will be discussed. The first is bootstrapping a worm program; and the second is three real worm programs: a simple, sequential worm which builds a tree of processors; a simple, parallel worm which does the same thing; and a more complex, sequential worm which can completely test and construct a map of a network. Each of these worms can be run from a processor connected by a link to the network. The program for this control processor, the root node, is also included for each example.

Bootstrapping a Processor

Transputer programs which are booted down a link are booted in two stages: first, a small bootstrap program is sent down the link, and second, this bootstrap program loads the main program which follows it. Usually, the bootstrap is generated automatically by the compiler and included with the program code. The bootstrap for a normal multiprocessor program must pass the code for the various processors to the correct destination processor and then begin executing its own program.

The bootstrap for a worm program, on the other hand, is somewhat simpler. Because the worm itself is a single-processor program, there is no need for a complex bootstrap program for ferrying code from processor to processor. The processor is booted simply by passing the compiled worm code (including the bootstrap) down one link, where it then begins running. The difficulty comes when a program tries to replicate itself. Where does the program to send down the new child's link come from? A worm has two choices: it can either keep a copy of itself (both program code and bootstrap) in memory or keep a copy only of the bootstrap in memory and somehow pass the program which it is executing down the link. In the first alternative, the bootstrap and worm must be stored in an array in memory; in the second alternative, only the bootstrap needs to be stored.

The first method, keeping a copy of the worm in an array, is very easy to program since the compiler automatically creates a file containing both the bootstrap and the worm program code. When booting a child, the worm will simply send its copy of the entire file, and then, when the child itself has woken up, send the copy again. The second time the child can input the copy into memory. Initially, of course, the root processor must get the worm code from somewhere, perhaps from the host computer. If a worm is booted in this way, the object code for the worm itself is created just like any other program with one processor. The outer PROGRAM fold should look like Fig. 7-4. Notice that no links are defined.

The basic problem with this simple approach is that the worm must allocate enough space to store a copy of itself. Doing this requires a large workspace and effectively doubles the size of the worm.

The second approach, having the worm copy its own executing program down the link, is more elegant but also more involved. Essentially, the worm loads

```
... SC worm
PLACED PAR
  PROCESSOR 0 T8
    worm()
```

Figure 7-4
Program level code for a worm

a small bootstrap routine separately, and this bootstrap in turn loads a new copy of the worm program and records the program and parameter locations in memory. This bootstrap must be explicitly written by the user. The bootstrap then runs the worm, passing the appropriate information to the worm to enable it to copy itself. Only the bootstrap is stored in the worm, but since this bootstrap is much smaller than the worm program itself, it does not take up as much space as with the first approach, keeping a copy of both the bootstrap and worm in memory.

The construction of a worm which copies its own executing program down a link relies heavily on special routines and programs available in the development system. (These routines are discussed in detail in the documentation for the development system[†].) In constructing the worm, we must first write a bootstrap which can read in the worm program itself and pass the appropriate parameters to the worm. Refer to Fig. 7-5 for the bootstrap used in the worm examples. This bootstrap is taken from an example in the development system. The channel down which the bootstrap is loaded is automatically passed to the bootstrap itself. This channel, the booting link `boot.link`, is the only argument the bootstrap has. Three parameters are associated with the worm program to be booted: the `work.-size`, the `code.size`, and the `entry.offset`. These parameters tell the loaded worm program how to begin executing properly. The parameters are read in by the bootstrap from the booting processor, and the total work space needed for the routine is calculated; the size of this array depends on the size of the compiled bootstrap itself. At this point, the `space` array is retyped and abbreviated. Next, the bootstrap reads the code to be executed and organizes the parameters associated with the subroutine call. These parameters can be passed to and returned from the called routine, and include an array pointer which allows the called routine to copy itself down a link. The bootstrap calls two `LOAD.VECTOR` routines which actually set up the parameters to be passed to the worm. The `KERNEL.RUN` command then begins executing the worm code read in. At this point the worm program itself is running. If the worm were to finish, control would be passed back to the bootstrap which could, because of the WHILE TRUE loop, read in a new block of object code and begin executing it.

---

†. INMOS Limited, *Transputer Development System*, (Hemel Hempstead, U.K.: Prentice Hall International (U.K.) Ltd., 1988).

```
PROC bootstrap (CHAN OF ANY from.boot)

  INT total.work.space:
  VAL INT num.of.parms IS 3:
  [10]INT data:
  [1636+20]BYTE space:
  INT work.size,code.size,entry.offset:
  WHILE TRUE
    SEQ
      from.boot ? [data FROM 0 FOR 3]
      code.size,entry.offset,work.size:=data[0],data[1],data[2]
      total.work.space := ((work.size+2)+num.of.parms)<<2

      []INT work.space RETYPES[space FROM 0 FOR
                                            total.work.space]:
      code IS [space FROM total.work.space FOR code.size]:
      SEQ
        from.boot ? code
        []INT parameters IS [work.space FROM work.size FOR
                                            (2+num.of.parms)]:
        SEQ
          []BYTE byte.data RETYPES data:
          LOAD.BYTE.VECTOR(parameters[1],byte.data)
          LOAD.BYTE.VECTOR(parameters[2],code)
          parameters[3] := data[0]
        KERNEL.RUN (code,entry.offset,work.space,num.of.parms)
:
```

Figure 7-5
A bootstrap program

The worm itself must contain the original bootstrap needed to boot a new processor down a link. The development system provides a utility, which, when applied to the compiled procedure, produces a file containing the object code of this bootstrap. This file is also included in the worm program.

A Simple, Sequential Worm

The first worm we will discuss is a simple, sequential worm which creates a tree of processors searched breadth-first. Each processor knows where its parents and children are, but does not attempt to find other neighbors. This worm is very simple to program but relatively slow in execution, and is suitable for initializing a network to run a farming program such as the one described in the chapter on processor farms. It should not be used to test networks of unknown reliability since no provision is made for error recovery.

A high-level view of the simple worm program is shown in Fig. 7-6. Each of the code folds is shown in subsequent figures as indicated on the fold line itself.

```
PROC worm([10]INT link.data,[]BYTE worm.code)
  ... Channels and Variables                              Fig. 7-8
  ... Proc poke.link(channel)                             Fig. 7-9
  SEQ
    ... clear values                                      Fig. 7-10
    ... find parent and read data                         Fig. 7-10
    ... wait for test command from parent                 Fig. 7-10
    SEQ link=0 FOR Number.of.Links
      IF
        link=parent.link
          SKIP
        TRUE
          SEQ
            ... test links                                Fig. 7-11
            ... if child found, then boot                 Fig. 7-11
    ... report number of children to parent               Fig. 7-11
    ... control loop                                      Fig. 7-12
    ... execute main program or return                    Fig. 7-13
:
```

Figure 7-6
High-level view of simple worm

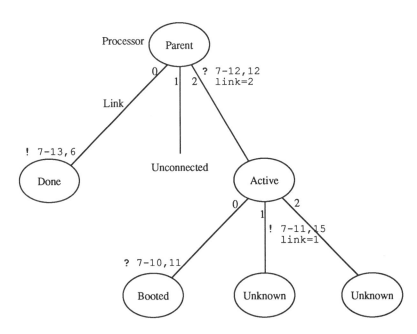

Figure 7-7
A portion of a network under investigation by the simple worm

One can reconstruct the entire program by inserting each figure into the appropriate fold. The worm proceeds by initializing itself, finding its parent, and then testing and booting any reset processors (children) connected to its links. When all of the links have been tested, the worm enters a passive mode, simply passing instructions and results up and down the tree. When all testing on any descendants is completed, the worm can begin executing a user program.

Figure 7-7 depicts a portion of a processor network at one point in the simple worm's execution. Six processors are shown together with their position in the worm program. The question or exclamation mark by a processor's link indicates whether the processor is doing (or trying to do) an input (question mark) or an output (exclamation point) on that link, and the numbers following the mark are the figure number and the line in the figure which the processor is executing. The links are ordered from left to right and their addresses indicated. By occasionally referring to this figure as you read through the text and code, you may find it easier to understand the overall worm program. In this figure, the parent node has already tested and booted all of its children. The first child is completely finished testing, the second link is not connected to any processor, and the second child is actively testing and booting its second child.

The worm program begins with a definition of the links (defined as CHAN OF ANY for simplicity and flexibility) and the parameters passed by the bootstrap. In Fig. 7-8 two libraries are included, one containing the object code of the boot-

```
#USE reinit
#USE "bootstrap_lib.tsr"
VAL boot.code    IS T8.bootstrap:
VAL INT start    IS  #80000000:
VAL INT Number.of.Links IS 4:
VAL INT LinkTimeOut IS 500:
[Number.of.Links]CHAN OF ANY link.in,link.out:
PLACE link.in AT Number.of.Links:
PLACE link.out AT 0:

boot.data                       IS [link.data FROM 0 FOR 3]:
INT parent.link                 IS link.data[3]:
INT Number.of.Children          IS link.data[4]:
[Number.of.Links]INT child.link IS [link.data FROM 5 FOR
                                                Number.of.Links]:
INT sum                         IS link.data[5+Number.of.Links]:
INT address,data:
[Number.of.Links]BOOL child:    --TRUE if child exists on link
BOOL Aborted:
VAL INT NotTried is -1:
```

Figure 7-8
Channel and parameter definitions for the simple worm

strap (created with a routine from the development system) and the second containing the time-out link input/output routines discussed later. The start address of the on-chip memory is defined together with the Number.of.Links on the processor and the time-out value for waiting on a tested link. Following this, the channels are defined and placed at their hardware addresses. The parameters used for booting a child, and which are returned to the bootstrap, are abbreviated conveniently. The parameters describing the worm code itself are included, as well as those for the parent link, the number of children, a list of the links off which the children are to be found, and the total number of children the worm has. Two variables (address and data) are used for testing links, and a boolean array called child is used to store a record of which links have been booted. An additional boolean for testing is included, and a parameter called NotTried is defined. Notice that some of these variables are used globally, rather than being passed as parameters in procedure calls. This is done to reduce the overall compiled code size.

In this simple, sequential worm, the actual link testing is done by the process poke.link shown in Fig. 7-9. The main part of the worm program will begin after this routine. The memory of an unbooted processor is tested in response to messages passed to it by poke.link. If a message string begins with a zero byte, the instruction is a write; a one byte indicates a read. The next word sent is the address, followed (for a write) by the data. An unbooted processor responds to a read command with the data stored at the address given. If no processor is connected to the link being tested, the link will hang up and the program will be unable to proceed. To prevent this, the development system provides a set of routines which perform an output, but which will time-out after a while if there is no response. These routines are called OutputorFail.t and InputorFail.t. A discussion of these routines can be found in the transputer development system documentation.

The simple worm proceeds with the link test by creating a 14-byte string which instructs the processor to do a write followed by a read. First OutputorFail.t and then InputorFail.t is called with the time-out parameter defined earlier. The boolean variable Aborted is set to TRUE or FALSE depending on whether the write and read succeeded (FALSE) or not (TRUE). The routine then returns control to the process which called it.

After the variables and the link-testing procedure are defined, the worm itself begins by clearing the count of children, the record of the children's existence, and the record of the children's locations. The address used to test processors connected to links is set and the testing data value cleared.

The next step in the worm program, shown in Fig. 7-10, is very important. To avoid trying to test its parent and experiencing the resulting complications, each processor must know where its parent is. The processor can find its parent by entering an ALT structure which attempts an input on each of its links (line 9). The link connected to the parent then carries a dummy value from the parent processor to the worm (sent in line 17, Fig. 7-11). The link over which the dummy value is passed is the parent link. At this point, the processor stops and waits for a further

```
PROC poke.link(CHAN OF ANY link.in,link.out)
  TIMER timer:
  INT time:
  VAL [4]BYTE byte.address RETYPES address:
  VAL [4]BYTE byte.data    RETYPES data:
  [14]BYTE s:
  SEQ
    s[0]:=(BYTE 0)                            --test sequence
    [s FROM 1 FOR 4]:= byte.address
    [s FROM 5 FOR 4]:= byte.data
    s[9]:=(BYTE 1)
    [s FROM 10 FOR 4]:= byte.address

    timer? time                               --check time
    OutputOrFail.t(link.out,s,timer,(time PLUS
                              LinkTimeOut),Aborted)
    IF
      Aborted                                 --no response
        SKIP
      TRUE                                    --response
        SEQ
          timer? time
          InputOrFail.t(link.in,[s FROM 0 FOR 4],timer,
            (time PLUS LinkTimeOut),Aborted)
:
```

Figure 7-9
Link-testing process for a simple worm

```
SEQ link=0 FOR Number.of.Links              --clear data
  SEQ
    child.link[link]:=NotTried
    child[link]:=FALSE
Number.of.Children:=0
data:=0
address:=start
ALT link=0 FOR Number.of.Links              --find parent
  link.in[link]? parent.link
    parent.link:=link
link.in[parent.link]?  data                 --command?
```

Figure 7-10
Initialization code for the simple worm

command from its parent. Since the worm is doing a sequential search, this wait prevents more than one processor at a time from testing its own links. When the processor does receive a command, it is ready to begin testing.

The testing proceeds sequentially on each link in turn in numerical order. Figure 7-11 shows the simple worm's code for testing and booting. As each of the links is tested in turn, the IF statement ignores the parent link and only tests the others. Testing a link is accomplished with the poke.link routine described earlier. If the output is not read in by the supposed processor at the other end of the link, Aborted will be set TRUE and the link will be ignored thereafter. If the output succeeds, Aborted is FALSE and the new processor can be booted. Note that if there is a processor at the other end of the link that has already been booted, it will not be listening to this link; it will either be waiting for a test command from its parent or running the control code in Fig. 7-12. In neither case will an already-booted processor input the test string output by the testing processor. Thus a link test can only succeed if there is an unbooted processor connected to the link.

Once the worm program has determined that there is a new (unbooted) processor off a particular link, the worm proceeds to boot it. In order for the worm to do this, the processor must output the bootstrap code, the parameters associated with the bootstrap, and the worm itself. Notice that these link outputs are matched exactly by the inputs found in the bootstrap. The last value output is a dummy. It simply tells the new worm which link is its parent. This dummy output corresponds directly to the input done by the ALT structure in Fig. 7-10.

When the new processor has been booted, the worm records the fact by setting the appropriate child element to TRUE, assigning the link value to the proper element of child.link, and adding the new processor to the total count by incrementing Number.of.Children. When all of the links of a given processor have been tested, the processor notifies its parent by passing the total number of the processor's new children back to the parent. At this point, the processor's children have been booted but have not yet tested their own links.

Once a processor has completely tested its links and possibly booted some of its own children, it is responsible for passing commands between its parent and its new children. The control code for doing this is shown in Fig. 7-12. This portion of the simple, sequential worm is really the most complex. As long as the processor has some unexplored descendant, it must continue to pass instructions to its children. When all of its descendants have tested their links, the exploration is done and the worm's task is finished. The worm always knows how many unexplored descendants it has since, after doing the testing, each descendant returns the count of its children just booted. This count is passed upward through the network to the controller.

The variable sum (in Fig. 7-12) keeps a running count of the number of each processor's descendants which have not yet tested their links. Initially, sum is set to the number of children that the worm has just booted and whose links still need to be tested. This sum is then passed to the processor's parent. The worm program

Chapter 7 — Worms

```
SEQ link=0 FOR Number.of.Links
  IF
    link=parent.link                             --parent link
      SKIP
    TRUE                                         --link to test
      CHAN OF ANY Link.under.Test IS link.out[link]:
      SEQ
        poke.link(link.in[link],Link.under.Test)
        IF
          Aborted                                --set by poke.link
            SKIP
          TRUE
            SEQ                                  --boot child
              Link.under.Test!   boot.code
              Link.under.Test!   boot.data
              Link.under.Test!   worm.code
              Link.under.Test!   0               --dummy output
              child.link[Number.of.Children]:= link
              Number.of.Children:=Number.of.Children+1
              child[link]:=TRUE
link.out[parent.link]! Number.of.Children        --tell parent
```

Figure 7-11
Test and boot code for the simple worm


```
sum:=Number.of.Children
WHILE sum>0                                      --test all descendants
  SEQ
    sum:=0
    link.in[parent.link]?  data                  --wait for command
    SEQ test=0 FOR Number.of.Children            --test children
      VAL INT test.link IS child.link[test]:
      IF
        child[test.link]
          SEQ
            link.out[test.link]! 0               --do test
            link.in[test.link] ? data            --descendants?
            IF
              data=0                             --no descendants
                child[test.link]:=FALSE          --no more tests
              TRUE                               --some descendants
                sum:=sum+data                    --count descendants
        TRUE                                     --don't test
          SKIP
    link.out[parent.link]! sum                   --descendant count
```

Figure 7-12
Control code for the simple worm

Parallel Programs for the Transputer

then enters a WHILE loop in which it remains until `sum` equals zero and the worm has no untested descendants. The `child` array is now used to keep a record of which children still have their own untested descendants. Those children are still active; the others can be ignored. Initially, none of the worm's children has tested its own links, the children are all active, and the array simply contains a record of each child that exists.

Once all of the links have been searched, the control loop proceeds by clearing `sum`. The controlling node initiates a new test by sending a command down to its active children. Each worm receives the test command from its parent. The worm passes the test command in turn down each link with an active child. Each active child will pass the command and wait for a response from each of its active children. The message will only stop at a processor with no children or a processor whose links have not been tested.

The worm program finds the links with active children by testing `child`. If `child` is TRUE, the child off that link is active, the command is passed, and a count is returned by the testing node. If the count returned is zero, there are no more unexplored descendants on that link and `child` is set FALSE (child is inactive) so that no more commands will be passed down the link. If the count is not zero, `sum` accumulates the count. When all of the links have been checked, the worm passes the total count up to the parent processor. Since the parent processor is running the same routine as its child, the parent in turn passes the total received from all its children up the tree until eventually the root node receives a total count of newly booted processors for the network. Thus, with a single command, an entire generation of descendants will be instructed to search its links, one processor at a time.

When a worm has no more untested children, its work is complete and the processor can proceed to its final task, executing a user program. This might be a farming program such as that illustrated in the chapter on processor farms. As a simple example of a user program following the worm, the network simply counts its members. Figure 7-13 shows the code necessary to do this. If a processor has no children, it sends to its parents a one, counting itself. A processor with children reads the counts passed by its children, sums them, increments the total, and passes it to its parent. The root node will then receive the complete count.

```
sum:=1                                          --count me
SEQ test=0 FOR Number.of.Children         --check all children
   SEQ
      link.in[child.link[test]]? data               --children?
      sum:=sum+data                          --sum descendants
link.out[parent.link]! sum                        --tell parent
```

Figure 7-13
User program following the simple worm

```
INT count:
SEQ
  SEQ                                         --initial boot
    out! T8.bootstrap
    out! [code.length,entry.offset,work.space.size];binary;0
  count:=1                                    --one alive
  WHILE count>0                               --while someone alive
    SEQ
      out! 0                                  --test next generation
      in? count                               --number new nodes
  in? count                                   --total nodes in tree
```

Figure 7-14
Root node control code for the simple worm

The root processor code necessary to start and run the simple, sequential worm discussed in this section is very simple and is shown in Fig. 7-14. The first worm is booted from a known link, and after a test command is sent to it, the first worm returns a count of the processors it in turn booted. Another test command is then sent, and the process is repeated as long as the count returned is greater than zero. When the count returned is zero, the exploration is finished and the controller inputs the total count of processors in the network.

This simple, sequential worm, when compiled with all compiler flags off, requires 760 bytes of code and work space and can test a 120-node network in less than 8 seconds using a 32 msec time-out for the link tests. The time required depends to a great extent on the time-out value used. Every test of an unconnected link, or of a link connected to a processor already booted, will time-out. Since timing-out happens frequently, and only one test proceeds at a time, the overall searching rate is quite slow.

A Simple, Parallel Worm

Running a parallel worm is a bit like driving an automobile without a steering wheel. Once you have started it is difficult to remain in control. Nonetheless, it is possible to create a simple, parallel worm program which runs very fast, is quite small, and is very simple to control. The parallel worm described here is similar to the simple, sequential worm in that it constructs a tree from a network of unbooted processors without attempting to discover anything about connected neighbors that are neither parent nor child. The parallel worm program is especially suitable for quickly booting processor farm applications on arbitrary networks.

The overall structure of the parallel worm is shown in Fig. 7-15. This structure is similar to the sequential version, but has three important differences. First, all of the links are tested in parallel; second, the link test uses the boot code itself rather than the poke.link process; and third, no control loop is necessary since

Worms   Chapter 7

```
PROC worm([10]INT link.data,[]BYTE worm.code)
  ... Channels and Variables                              Fig. 7-17
  SEQ
    ... clear values                                      Fig. 7-18
    ... find parent                                       Fig. 7-18
    PAR link=0 FOR Number.of.Links
      IF
        link=parent.link
          SKIP
        TRUE
          SEQ
            ... test links                                Fig. 7-19
            ... if child found, then boot                 Fig. 7-19
    ... count children                                    Fig. 7-19
    INT sum,data:
    ... execute main program or return                    Fig. 7-13
:
```

Figure 7-15
High-level view of the parallel worm

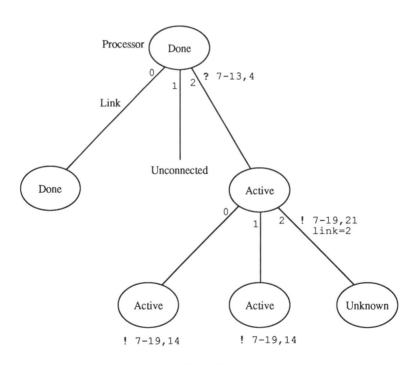

Figure 7-16
A portion of a network under investigation by the parallel worm

Parallel Programs for the Transputer

each processor is running independently and has no need to report to its parent or to the root node. This link testing approach lacks finesse, but using it simplifies the control procedures for multiple worms running at the same time. The sequential worm could also use the same link testing approach and would then run somewhat faster.

Figure 7-16 depicts the same network shown in Fig. 7-7, but, in this case, the network is being explored by the parallel worm. The link and figure notations in this figure are similar to those in Fig. 7-7. This time, the network has three active processors investigating their links at the same time. Two other processors are done, and one is yet unexplored. The children of the two active processors at the bottom of the tree are not shown. Once again, you will find it very helpful to refer to Fig. 7-16 as you read through the text and programs.

The methods used for booting the parallel worm are identical to those described earlier for the sequential worm. Figure 7-17 shows the variables and channels used. The only real difference between this and Fig. 7-8 (showing the definitions for the sequential worm) is the omission of the sum, address, and data values, and the start address. The Aborted flag used for the Outputor-Fail.t call is declared later in the parallel worm.

After it is booted, the parallel worm begins running by clearing the children count and the list of child.link, and by setting the child variables to TRUE (Fig. 7-18). Initially, the child variable is used to keep track of links which might be connected to the parent, and does not indicate whether a child on a link is booted. Later, child will indicate whether a child on a link is booted, and will be initialized to FALSE.

In order to find its parent, a processor running the parallel worm cannot simply construct an ALT and listen on all of its channels as the sequential worm does, because there may be many active processors running at any given moment and one of them may be trying to test *this* particular processor. If an ALT reads the test input from a processor other than its parent, it might become confused. So that the worm can distinguish between the test of another processor and a message from its parent, the worm inputs the first byte and examines it. A boot packet cannot begin with a zero byte because a zero indicates a memory write down a link rather than program code. Therefore, if the parent passes a zero byte, the child can readily distinguish the parent from other testing processors. To accomplish this, the parallel worm sets a comparison byte to 1 and enters a WHILE loop, which repeatedly does an ALT input on its channels and waits for a zero byte on a link. If a byte which is not a zero arrives, it must be the first byte of a boot packet, and the entire packet must be disregarded since it is not from the parent processor. To prevent subsequent inputs on the same link, the input for the ALT structure uses a guard, the child boolean. When a nonzero byte does arrive on a link which is not connected to the parent, the boolean is set to FALSE and the data on that link are ignored. Eventually, the parent will get its zero byte through, the parent link is determined,

```
#USE reinit
#USE "bootstrap_lib.tsr"
VAL boot.code     IS T8.bootstrap:
VAL INT Number.of.Links IS 4:
VAL INT LinkTimeOut IS 500:

[Number.of.Links]CHAN OF ANY link.in,link.out:
PLACE link.out AT 0:
PLACE link.in AT Number.of.Links:

boot.data                       IS [link.data FROM 0 FOR 3]:
INT parent.link                 IS link.data[3]:
INT Number.of.Children          IS link.data[4]:
[Number.of.Links]INT child.link IS [link.data FROM 5 FOR
                                    Number.of.Links]:
[Number.of.Links]BOOL child:

VAL INT NotTried       IS -1:
```

Figure 7-17
Variables and parameters for the parallel worm

```
SEQ
  SEQ link=0 FOR Number.of.Links           --clear everything
    SEQ
      child.link[link]:=NotTried
      child[link]:=TRUE
  Number.of.Children:=0
  BYTE byte:
  SEQ
    byte:= (BYTE 1)
    WHILE byte <> (BYTE 0)                 --wait for parent
      ALT link=0 FOR Number.of.Links       --wait for input
        child[link] & link.in[link]? byte
          IF
            byte = (BYTE 0)                          --parent
              parent.link:=link
            TRUE                                --somebody else
              child[link]:=FALSE
  SEQ link=0 FOR Number.of.Links                      --clear
    child[link]:=FALSE
```

Figure 7-18
Start-up code for the parallel worm

and the WHILE loop is escaped. The `child` variable is then set to FALSE and serves as a record of processors booted, just as it does for the sequential worm.

Once a processor's parent link has been determined, the worm is ready to test the rest of the links (Fig. 7-19). This testing operation can proceed in parallel on each of the links. Since most of the time in the test code is spent doing link input or output, a parallel structure should work very efficiently.

The parallel worm proceeds with testing links by constructing a PAR structure for each of them. Each parallel process must create its own variables and abbreviations. `Aborted` is defined as the time-out boolean. The link associated with each process is abbreviated to distinguish it from the other links, as are the variables which keep track of which link has a child booted on it.

If a process' link is the parent link, it does not need to be tested. If the link is not the parent link, testing must proceed. The testing is done very simply, by the program trying to output the first byte array necessary to boot a processor. If this attempt fails, the test will time-out and `Aborted` will be set to TRUE. If, on the other hand, the attempt succeeds, the worm program knows that there is a child to be booted off the link and it proceeds to complete the boot process by passing the boot data and worm code. The last byte sent to the new child from the processor must be a zero byte which informs the new child which link is connected to its parent. This output (the zero byte) corresponds directly to the ALT input done in Fig. 7-18. Finally, once the child is booted, the `child` variable must be set to TRUE and the link recorded.

When all of the processor's links have been tested and the available children booted, the worm should consolidate the information from each of the parallel processes so it can easily pass information to its children. The worm readily accomplishes this consolidation by testing `child` to see which links have children, copying the link numbers down into a consecutive list, and counting the number of children booted.

At this point the worm is finished and the program that follows the worm can begin executing. As with the sequential worm, we use as an example of an application program a simple procedure for counting the number of nodes in the network. Each processor checks its children and inputs the number of descendants each child has. These are then summed and passed to the parent. The code to do this is identical to that of the sequential worm (with the added definition of the integer variables `sum` and `data`) and is shown in Fig. 7-13.

The root node control program which runs the parallel worm is even simpler than that for the sequential worm. All that is necessary is for the controller to boot the first processor and wait. At the end of the program, the count of the number of processors can be read in. The control code is shown in Fig. 7-20.

The structure developed by a parallel worm is interesting to examine. Figure 7-21 shows the tree constructed by the parallel worm on a 20-node toroidal network with the root connected to the top link of the upper right node. Although the tree

```
PAR link=0 FOR Number.of.Links              --test all links
  BOOL Aborted:
  CHAN OF ANY Link.under.Test IS link.out[link]:
  child.link IS child.link[link]:
  child.exist IS child[link]:
  IF
    link=parent.link                        --skip parent link
      SKIP
    TRUE                                    --test this link
      TIMER timer:
      INT time:
      SEQ
        timer? time
        OutputOrFail.t(Link.under.Test,boot.code,timer,
                       (time PLUS LinkTimeOut),Aborted)
        IF
          Aborted                           --no answer
            SKIP
          TRUE                              --answer so boot
            SEQ
              Link.under.Test!    boot.data
              Link.under.Test!    worm.code
              Link.under.Test!    (BYTE 0)
              child.link :=link
              child.exist:=TRUE
SEQ link=0 FOR Number.of.Links              --make a list of children
  IF
    child[link]
      SEQ
        child.link[Number.of.Children]:=child.link[link]
        Number.of.Children:=Number.of.Children+1
    TRUE
      SKIP
```

Figure 7-19
Test and boot code for the parallel worm

```
INT count:
SEQ
  out! T8.bootstrap                         --start first one
  out! [code.length,entry.offset,work.space.size]
  out! binary;(BYTE 0)
  in? count                                 --wait for reply
```

Figure 7-20
Root node control code for the parallel worm

Chapter 7   Worms

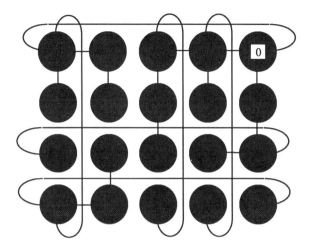

Figure 7-21
Tree constructed by the parallel worm

is not identical with that created by a breadth-first search, it is roughly similar and no deeper. The same tree is constructed every time the worm is executed.

The parallel worm that has been described in this section is very small and very fast. The total code and work space requirement is 715 bytes. On a 120-node system, this worm finishes exploring in less than 0.04 seconds using a link time-out value of 32 msecs.

A Robust, Exploratory Worm

A common use for worms is exploring and testing unknown networks of unknown reliability. In this case, it is imperative that the worm be able to recover from hardware faults and that a report be passed to the root node describing the network and any failures found. This requirement makes the worm much more complicated and the exploration strategy more involved. An exploratory worm in an unknown network of unknown reliability must also make a complete report of *all* the interconnections in the network, not just those of parent and child as in our two previous examples.

This exploratory worm is sequential and can search a network in any way the root node desires. Since every node tested requires a separate command, the ordering of node tests is arbitrary. For this example, the root node will do a breadth-first search, and after every test, a complete report of the findings of the node tested will be returned to the worm controller on the root node. If for some reason a communication should fail, a failure can be reported as well.

A truly robust worm, one which can recover from any error whatsoever, is difficult to construct. After all, in the worst case, the power could fail for the entire system, leaving the user with a blank screen. Less drastic errors, however, do occur; one of the most difficult to detect is that of intermittent failures on a link or processor. Special efforts must be made by the worm to recover from and report such failures. The exploratory worm described here allows the user to repeatedly test a single node in the hope of being able to discover such intermittent failures.

In Fig. 7-22 the high-level structure of the exploratory worm is listed. The exploratory worm uses the same routines for booting that the simple sequential and parallel worms did. The parameters are passed in by the bootstrap at the beginning and the channels and variables defined together with two input and two output routines. Of these routines, two are used for testing and the others for passing integer messages up and down the network tree created by the worm. The poke routines write to the link, and the peek routines read from the link.

The diagram shown in Fig. 7-23 to illustrate the exploratory worm program is slightly different from the earlier diagrams. This time the active node is connected to another child and has only two children of its own on the bottom layer. The links are ordered as shown rather than from left to right as before. In this diagram, the top processor has booted both of its children, one of which is now active and investigating its own links and is waiting for something to happen on any link. At the moment depicted in the diagram, the active node has just sent a test message to the node on the left and is receiving a reply from it.

After the initial procedure definitions, the exploratory worm begins executing. Its first actions are to clear the pertinent variables and find its parent. Once these have been done, the worm enters a WHILE loop in which it remains until instructed by a kill command to exit. The WHILE loop consists of an ALT input on

```
PROC worm([10]INT link.data,[]BYTE worm.code)
    ... Channels and Variables                          Fig. 7-24
    ... Proc Poke.int (channel,int)                     Fig. 7-25
    ... Proc Poke.link(channel)                         Fig. 7-25
    ... Proc Peek.link(channel)                         Fig. 7-25
    ... Proc Peek.int (channel,int)                     Fig. 7-25
    SEQ
        ... clear values                                Fig. 7-26
        ... find parent and read data                   Fig. 7-26
        WHILE (status <> kill)
            ALT link=0 FOR Number.of.Links
                link.in[link]? status
                    IF
                        ... message from parent         Fig. 7-28
                        ... message from child          Fig. 7-27
                        ... message from unknown        Fig. 7-27
:
```

Figure 7-22
High-level code for the exploratory worm

Chapter 7   Worms

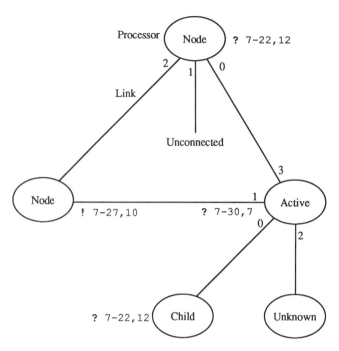

Figure 7-23
A portion of a network under investigation by the exploratory worm

every link, followed by the appropriate action. If the input is from the parent processor, the worm must respond, usually with a test and boot procedure. If the input is from a child, the worm simply passes the data back up the tree to the parent. If the input is from an unknown processor, the worm assumes it is being tested and responds appropriately.

The basic action of the worm, then, is to respond to link inputs. Inputs from parents are commands to be acted upon; inputs from children are not acted upon and are simply passed along; other link inputs are assumed to be other processors trying to test the link. These tests are replied to and recorded, but no action is taken by the worm.

Figure 7-24 lists the constants, and the channel, variable, and library definitions for the exploratory worm; most of them are the same as those of the two previous worms. A few additional constants and variables are added: the variable me uniquely identifies each processor; a list of processors (proc.list) connected to each link is maintained with an extra boolean for link input; the action of the worm is controlled by the status variable; and the address and data values are much as before. A list of constants representing the various failure modes of the processors or the responses to the link tests is also defined.

```
#USE reinit
#USE "bootstrap_lib.tsr"
VAL boot.code      IS T8.bootstrap:
VAL INT start      IS  #80000000:
VAL INT Number.of.Links IS 4:
VAL INT LinkTimeOut IS 500:

[Number.of.Links]CHAN OF ANY link.in,link.out:
PLACE link.in AT Number.of.Links:
PLACE link.out AT 0:

boot.data                        IS [link.data FROM 0 FOR 3]:
INT Parent.Link                  IS link.data[3]:
INT Number.of.Children           IS link.data[4]:
[Number.of.Links]INT child.link  IS [link.data FROM 5 FOR
                                     Number.of.Links]:
INT me                           IS link.data[5+Number.of.Links]:

[Number.of.Links]BOOL child:
[Number.of.Links]INT  proc.list:
BOOL Aborted,In.Aborted:
TIMER timer:
INT address,data,status:

VAL INT TimedOut       IS -1:
VAL INT NotTried       IS -1:
VAL INT Kill           IS -1:
VAL INT NewProcessor   IS -2:
VAL INT BadProcessor   IS -3:
VAL INT NoProcessor    IS -4:
VAL INT LiveProcessor  IS -5:
VAL INT MemoryError    IS -6:
VAL INT LinkError      IS -7:
VAL INT SelfConnected  IS -8:
```

Figure 7-24
Variables and parameters for the exploratory worm

```
PROC peek.link(CHAN OF ANY link)
  INT time:
  [10]BYTE s:
  SEQ
    timer? time
    InputOrFail.t(link,s,timer,(time PLUS
                                LinkTimeOut),In.Aborted)
:
```

Figure 7-25
Link input/output routines for the exploratory worm

```
PROC peek.int(CHAN OF ANY link,INT result)
  INT time:
  [4]BYTE s RETYPES result:
  SEQ
    timer? time
    InputOrFail.t(link,s,timer,(time PLUS LinkTimeOut),Aborted)
:

PROC poke.int(CHAN OF ANY link,VAL INT value)
  INT time:
  VAL [4]BYTE s RETYPES value:
  SEQ
    timer? time
    OutputOrFail.t(link,s,timer,
                        (time PLUS LinkTimeOut),Aborted)
:
PROC poke.link(CHAN OF ANY link)
  INT time:
  VAL [4]BYTE byte.address RETYPES address:
  VAL [4]BYTE byte.data    RETYPES data:
  [14]BYTE s:
  SEQ
    s[0]:=(BYTE 0)
    [s FROM 1 FOR 4]:=byte.address
    [s FROM 5 FOR 4]:=byte.data
    s[9]:=(BYTE 1)
    [s FROM 10 FOR 4]:=byte.address

    timer? time
    OutputOrFail.t(link,s,timer,
                        (time PLUS LinkTimeOut),Aborted)
:
```

Figure 7-25 (cont.)
Link input/output routines for the exploratory worm

```
SEQ i=0 FOR Number.of.Links
  SEQ
    proc.list[i]:=NotTried
    child[i]:=FALSE
Number.of.Children:=0
status             :=0
ALT i=0 FOR Number.of.Links
  link.in[i]? me;proc.list[i]
    Parent.Link:=i
```

Figure 7-26
Variable initialization for the exploratory worm

The exploratory worm uses four link input and output routines (Fig. 7-25). These routines are all similar. The first process, peek.link, tries to read in a 10-byte string, using InputorFail.t. If it fails to input the string, a boolean, In.Aborted, is set. The second process, peek.int, tries to read an integer and, upon failure, sets the boolean Aborted to TRUE. The third process, poke.int, attempts to output an integer down a link using the Outputor-Fail.t process. The integer passed must be retyped into a byte string. If the output fails, Aborted is set to TRUE; otherwise it is set to FALSE. The fourth process, poke.link, does the same thing, but with a test string instead of a simple integer. The string is a write-then-read memory test with a beginning byte of zero, an address, and a data value, followed by a read address command. All of these are compacted into a 14-byte string. Once again, if the string cannot be output, Aborted is set to TRUE.

Notice that these peek and poke routines are complementary. Process poke.link is the complement of peek.int followed by peek.link; poke.link outputs 14 bytes; and peek.int followed by peek.link inputs 14 bytes. Thus the peek routines can be used to "swallow" tests from other processors. The two peek and poke integer routines are useful for passing integer messages from one processor to another.

Following the definition of the variables and procedures, the exploratory worm can begin executing (Fig. 7-26). Its first action clears the variables which record the processor list and the record of children booted in child. The child count and the status parameter are set to zero as well. Once these preliminaries are out of the way, the worm can search for its parent. This search is done in the usual way, with an ALT waiting on all of the link inputs. The parent processor that booted this worm must pass a value to the child worm to enable it to find the parent link. The first value read in from the parent, however, is not just a dummy as it was for the simple sequential and parallel worms, but the value me, which is the node's particular identification number. A second value, the parent processor's identification number, follows and is stored in the list of processors.

Following this initial interaction, the worm is ready to begin active duty. It enters a WHILE loop controlled with the status value (Fig. 7-22). The program simply waits in a loop for an integer input on any link. If the link providing the input is the parent link, the worm considers the input to be a command; if the input is from a child link, the data is passed to the parent; and if the input is from any other link, the processor concludes that it is being tested by another processor. After each input has been handled, and unless the kill command was sent, the worm returns to waiting for another input.

The responses to the last two possibilities (input from any nonparent link) will be described first and are shown in Fig. 7-27 together with a repetition of the IF statement and the parent link comparison. If the link on which the input was received is neither a parent link nor a child link, no response to the parent is expected. However, if a child is passing the data (child[link] is TRUE), the data must

```
IF
  link=Parent.Link                              --parent
    ... message from parent                     Fig. 7-28

  child[link]                                   --child
    link.out[Parent.Link]! status
  TRUE                                          --unknown
    SEQ
      peek.link(link.in[link])                  --read remainder
      link.out[link]! me                        --reply
      link.in[link]?  proc.list[link]
```

Figure 7-27
Response to child or unknown input for the exploratory worm

```
link=Parent.Link
  IF
    status=me                                   --message for me
      INT node:
      SEQ
        link.in[link]? node                     --node count
        SEQ
          SEQ link.under.test = 0 FOR Number.of.Links
            CHAN OF ANY Test.Link IS link.out[link.under.test]:
            IF
              proc.list[link.under.test]=NotTried    --test
                INT input.status,word:
                SEQ
                  PAR
                    ... output link test        Fig. 7-29
                    ... input link response     Fig. 7-30
                    ... act on test results     Fig. 7-31
              TRUE                              --don't test
                SKIP
          link.out[Parent.Link]! node;proc.list --return data
    TRUE
      SEQ test=0 FOR Number.of.Children  --message for children
        poke.int(link.out[child.link[test]],status)
```

Figure 7-28
Response to command from root node for the exploratory worm

in turn be passed on to the parent by an output on the parent link. Since the status variable is being used to store the initial link input, the data must not be equivalent to a kill command (which causes the worm to stop). If the input is not from a child or a parent, it is assumed to be from some other processor, and the only thing this other processor could be doing is testing one of its links. To avoid hanging up the link, the program must read in the entire test string (14 bytes). The first 4 bytes have already been read in with the ALT input to status and the remaining 10 bytes can be read in with peek.link. Once the test string has been read, the testing processor will expect a reply because the test string includes a four-byte read after the write. The worm responds with the value me, informing the remote worm of the identity of its new neighbor. The remote worm, discovering that its test subject is already alive (because the me response is different from the data value it tried to write over the link), replies in turn with its identity, which is stored away in the processor list.

The most involved work for the worm comes with a command from its parent. Commands received on the parent link have been passed down the tree from the control node. The code the worm uses for responding to a command from the controller is shown in Fig. 7-28. This command may be either for the processor itself or for one of its descendants, with the two cases distinguished by the status value read in. If status equals the processor's identity value, me, the message is for the processor. If it is not, the command message is for one of the processor's children, and the worm program passes the message on to all its children with poke.int calls. Notice that this procedure amounts to a broadcast from the controlling root node to every node in the tree except for the descendants of the node for whom the message is intended. This communication time is wasted for all the processors for whom the message is not intended, but this is unimportant since these other processors are simply waiting for an interaction anyway.

If the command is intended for the local processor, the worm reads a second value and stores it in the variable node. This variable is the count of the number of processors booted so far in the exploration. Since the processors are numbered in the order in which they are booted, node also represents the number of the next processor to be booted. (The first node booted is zero.) Since node is not equal to any booted node's identification number, processors above the local one will pass it on just as they did status earlier.

After receiving the node value, the worm progresses into its test state. In this state, each of the links is examined in order. The link being tested is abbreviated and, if this link has not been used before, the test proceeds. The parent link and any neighbors who have tested a link earlier will already have data in proc.list and will therefore be skipped. The link test proceeds in two stages: first a memory read/write is sent down the link, and at the same time, listened to on all other links; and then the message returned is analyzed and acted on.

There are basically four things that can happen on a good network at this point: the link being tested may be unconnected, connected to an unbooted proces-

```
SEQ
  address:=start
  data:=NewProcessor
  poke.link(Test.Link)                              --send test
```

Figure 7-29
Link test output for the exploratory worm

```
INT time:
SEQ
  timer? time
  ALT                             --wait for reply or time-out
    ALT link.returned=0 FOR Number.of.Links
      CHAN OF ANY Return.Link IS link.in[link.returned]:
      Return.Link? word                             --reply
      IF
        link.returned<>link.under.test              --me
          SEQ
            peek.link(Return.Link)         --read test
            proc.list[link.returned]   :=me
            input.status:=SelfConnected
        word=NewProcessor                           --new proc
          input.status:=NewProcessor
        TRUE                                        --booted proc
          input.status:=LiveProcessor
    timer?  AFTER time PLUS LinkTimeOut    --time-out
      input.status:=TimedOut
```

Figure 7-30
Link test input for the exploratory worm

sor, connected to a booted processor, or connected to the same processor as is doing the test (connected to itself). In the last case, the link input must execute in parallel with the output to successfully respond. For this, a PAR structure is used, with the output on one side and the input on the other side.

The output code for the exploratory worm's link test is very simple and is shown in Fig. 7-29. The address of the memory test is set, the variable `data` is set to `NewProcessor` (a number which cannot be a node identification number), and the link is tested with `poke.link`. The input code in the second half of the PAR is found in Fig. 7-30. An ALT listens on every channel, with a time-out. Variable `input.status` is used to record the result. If no input occurs, the input times out and `input.status` is set accordingly. If an input *is* received, but not on the same channel which output the test, then the processor must have a link connected to itself. This fact and the link connections are then recorded. If an input is received on the *same* channel, then the worm concludes that there is another processor on the other end. If the value of the input received is the same as that sent (`NewProcessor`), the worm assumes that the processor has not been booted before. If the value of the input received is different than the one sent, the worm assumes that another living processor has responded with its own identity, and sets the `input.status` accordingly.

Once the test results have been obtained, the worm acts on those results as shown in Fig. 7-31. First, if the output and the input timed out, the worm assumes that there is no processor on the other end of the link. But if the output timed out and the input did not, something strange happened, and `proc.state` records the situation as a bad processor. Likewise, if the output was all right but the input was not, a bad processor is again recorded. This second problem could also indicate an intermittent link.

If the status indicates a living processor, the worm program must return its identity to the remote processor and store the remote processor's identity in `proc.state`. This exchange corresponds with the message-from-unknown link response found for the initial ALT input as shown in Fig. 7-27. The unknown processor's input of a test string, response output, and subsequent input corresponds with the worm's test output, response input, and identity output.

If `input.status` indicates that a new processor has been found, the new processor is booted. The link down which the new processor is booted is recorded by the worm, and the count of children is incremented. The worm then passes the new node value and its own identification number to the new child. The new node value becomes that child's identification number. This identification number output corresponds with the original ALT input immediately following the clear code at the start-up of a worm (Fig. 7-26). The new processor connection is recorded in `proc.state`, `child` is marked TRUE, and the node count is incremented.

It is also possible, after the initial link test is completed and a new processor has been found, to do a more intensive memory test of the new processor rather than just booting it. Using the read/write commands on the link, the worm can test the

```
INT proc.state IS proc.list[link.under.test]:
IF
  Aborted                                           --no output
    IF
      input.status = TimedOut              --no output no reply
        proc.state:=NoProcessor
      TRUE                                    --no output, reply
        proc.state:=BadProcessor
  TRUE                                                  --output
    IF
      input.status = TimedOut                         --no reply
        proc.state:= BadProcessor
      input.status = NewProcessor                        --boot
        SEQ
          Test.Link!   boot.code
          Test.Link!   boot.data
          Test.Link!   worm.code
          child.link[Number.of.Children]:= link.under.test
          Number.of.Children:=Number.of.Children+1
          Test.Link!  node;me
          proc.state:= node
          child[link.under.test]:=TRUE
          node:=node+1
      input.status = LiveProcessor                       --node
        SEQ
          Test.Link! me
          proc.state:=word
      TRUE                                        --self-conected
        proc.state:=me
```

Figure 7-31
Action on link test results for the exploratory worm

entire memory space of the processor to discover its size and whether any bad memory locations exist in the new processor's memory. These tests can conveniently be done just before the boot procedure. If a memory error is found, the processor state can be set to MemoryError and the link ignored. The memory-test code is not included in this worm program.

After all of the links have been checked, the worm returns to the code shown in Fig. 7-28. The link information and node count acquired are passed to the parent processor, and the worm goes back to waiting for a new input or command as shown in Fig. 7-26. Note that a second test command to the worm will invoke the same reply; the data acquired in the first set of tests will simply be returned.

If a kill command is sent by the controller, the command will be broadcast to every processor in the network since the command is different from any processor identification number. When a kill command is received, the worm will take no further action, but will simply escape the WHILE loop and return to the bootstrap.

The root node control code for this robust exploratory worm is somewhat more complex than that of the previous two examples, but is still fairly simple. Figure 7-32 shows the listing for the root node control code in the exploratory worm. After the first processor is booted, the node count is set to one and the current node under investigation is set to zero. The new worm's identity (zero) and the root node's own identity (here set arbitrarily to -33) are exchanged to complete the initial handshake. The controller then continues into a WHILE loop in which it will remain until every processor has been instructed to test itself. Once the controller is in the WHILE loop, the test command for the current node and the node count

```
VAL Number.of.Links IS 4:
VAL Max.Procs       IS 512:
VAL Kill            IS -3:
[Max.Procs][Number.of.Links]INT list:
INT node,node.count:
SEQ
  out! T8.bootstrap                        --boot first node
  out! [code.length,entry.offset,work.space.size];binary
  node:=0
  node.count:=1
  out! node; (-33)
  WHILE node.count <> node                 --test until none left
    SEQ
      out! node;node.count                         --send test
      in? node.count;list[node]          --reply, number & data
      node:=node+1                                 --next node
  out! Kill                                             --done
```

Figure 7-32
Root node control code for the exploratory worm

are sent out. The worm does its test and responds with the current node count and a list of its connections. The next processor is then instructed to test itself by passing its identity and the latest count of processors. This process continues until the number of processors tested equals the number loaded; all testing is then complete and the controller passes a kill command to the worm.

The robust exploratory worm, although much more complex than the simple sequential one, is also much faster since it does not time-out when communicating with nonparent or nonchild neighbors. It searches a 120-node network in less than 0.22 seconds. The worm itself occupies 1269 bytes of code and work space.

The very simple control routine described here does not make any attempt to deal with networks which do not behave properly. Improper behavior usually manifests itself to the controller by a complete lack of response to the controller's test commands. This situation can be dealt with by a time-out ALT on the data input. If such a time-out does occur, the controller can then find out where the error occurred by tracing back down the tree to the processor which was supposed to be tested. The root node traces the tree by retesting each processor in the path to the original processor under test. The first processor which does not respond is the one at fault.

It is not possible, with the worm as programmed here, to determine with complete certainty which link failed. If a link failure occurred as a command was being passed *down* the tree, the parent processor will time-out and resume duty. The parent processor can then be polled correctly and the error determined to be either the child processor or the link between the parent and child. If the failure occurs as a response is being propagated *up* the tree, the child processor will again fail to respond when a second inquiry is made. In this case, the parent processor could also be at fault.

An important concern with worm programs is their code size. An exploring worm that can run on any transputer network, regardless of the memory at each node, is especially useful. To have this capability, the worm code itself must be able to fit into the on-chip memory in the transputer. All three of the examples described in this chapter require less than 2048 bytes of memory and will run on any transputer with at least that much on-chip memory. A memory-testing routine following the worm can then effectively test all the off-chip memory (if any), providing a more complete system test.

In Summary

Worm programs are one of the most interesting kinds of programs which can be executed on a network of parallel processors. These programs can run on networks of arbitrary size and interconnection and are completely portable between networks of different structure. Worm routines are most useful for loading programs on arbitrary networks and for testing and debugging networks of unknown reliability and composition.

Worms programs can search networks in either breadth-first or depth-first order. A breadth-first search is slightly more complex than a depth-first search, but results in a wider and shorter network with a smaller interprocessor communication distance between the controller and the bottom of the tree.

Three worm programs are presented, two which are useful for loading application programs, and one which is capable of exploring unknown networks. Of the first two worms, the simple worm is slower but its search is completely controllable, while the parallel worm is very small and fast, but can create unpredictable networks. The exploratory worm is the largest and is capable of testing networks with faulty processors and interconnections.

Chapter 8

Real-Time Processing

Real-time processing is a natural application for small networks of parallel processors. A real-time processor is one which must respond to external events in the outside world within a fixed period of time, often before the next external event occurs. These events can occur at either a random or a fixed rate. Transputer systems are an excellent computing tool for real-time processing because they can be readily structured to meet the computing demands of an application which must interact with the outside world. Doing real-time processing often requires irregular computing structures, and the ability to easily structure and program a computer to meet the needs of a peculiar task provides an inherent advantage for transputer-based parallel computers.

Because of their multitasking capabilities, transputer systems offer additional advantages for the implementation of real-time systems. Real-time processing requires that an event in the real world be communicated to a computing system, a response be determined by the computer, and an action taken. Frequently, additional tasks in real-time systems execute at low priority when no external events interrupt the computer. When interrupts do occur, they are handled at a high priority so that the system responds as quickly as possible to events in the real world. The creation of and communication between tasks running in parallel are thus features of real-time systems. This communication between processes and the external world is very naturally modeled by occam processes and channels, and very naturally implemented by transputer microprocessors.

Minimizing hardware size and complexity is also an important goal in the design of real-time systems. Because transputer links provide an integrated solution to communication requirements within a computer, parallel computers based on transputers can provide a very simple and compact solution to real-time processing needs.

Not only is it important that real-time systems execute programs efficiently, but also that they allow for the immediate input of external events. If the system is busy processing previous events, it is difficult for it to input events without delay. If events occur, on average, more frequently than the time required to process them, a real-time system will fall behind and be unable to respond to new events immediately.

Whether events occur at fixed intervals or randomly, a real-time system must be able to process the events faster than the average rate at which the events occur, and do so without delaying the input of any of the interrupting events. In a real-time system responding to randomly occurring external events, the system will occasionally have to deal with sudden flurries of events which occur faster than the system can process them.

In this chapter are presented five programs which receive external interrupts and compute responses. The effectiveness of each program is found by measuring the average amount of time during which the program is unable to accept external events. An effective interrupt handler will respond to every event with minimal delay, while a less effective program will often experience delays between the time an event occurs and the time it is input.

Interrupt Handlers

A Simple Handler

The simplest possible interrupt handler is diagrammed in Fig. 8-1a and listed in Fig. 8-1b. This program is used as the baseline program against which any improved program is compared. In the routine, an external event sends a signal to the interrupt handler, which simply inputs the signal and computes the response. In Fig. 8-1b, the signal is input on channel in and stored in value data, and the response is computed with process handler. The response computation is application-specific; thus no example is listed here. The most important thing to note about this simple handler is that while it is computing a result it is completely unavailable to input a second signal. Any external event which attempts to signal the interrupt handler will either have to wait or abandon the attempt.

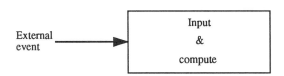

Figure 8-1a
A very simple interrupt handler

```
INT data:
WHILE TRUE
  SEQ
    in? data                                                    --input
    handler(data)                                               --work
```

Figure 8-1b
Code listing for a very simple interrupt handler

Chapter 8 — Real-Time Processing

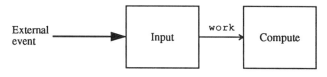

Figure 8-2a
A parallel interrupt handler

```
CHAN OF INT work:
PRI PAR
  INT data:                                                 --input
  WHILE TRUE
    SEQ
      in? data
      work! data

  INT data:                                                 --work
  WHILE TRUE
    SEQ
      work? data
      handler(data)
```

Figure 8-2b
Code listing for a parallel interrupt handler

A Buffered Handler

By adding a second parallel process to the interrupt handler, we can vastly improve it. The first process can perform the actual input while the second process computes the response (Fig. 8-2a). If the input process is run at a higher priority than the computing process, an external event can signal the handler even if the handler is computing a response from a previous signal. Thus the parallel handler essentially creates a high-priority buffer of size one in which to store interrupt events. The code for this interrupt handler is listed in Fig. 8-2b.

Although the parallel interrupt handler will be much more responsive than the simple input handler to external events, it will still fail if several events occur in a time shorter than the time needed to process them. The first event will be passed by the channel work to the computing process. Since the input process is running at a high priority, it can immediately respond to a second external event. However, the second event cannot be immediately passed along to the computing process, since the computing process is still busy with the first event. If a third event occurs, the input process will then be unable to respond.

Obviously, if the interrupt handler must respond to events occurring in quick succession, it needs additional buffering of some sort in which to store events oc-

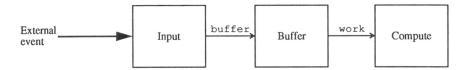

Figure 8-3a
A simple buffered interrupt handler

curring too rapidly to be processed as they occur. Since this buffer must be of finite size, in the worst case a sufficiently large number of events occurring in a short enough interval will always cause the handler to either delay the event inputs or lose the events entirely. The size of the buffer needed will depend on the relative frequency and distribution of the external events and the processing time needed to compute a response.

A buffered interrupt handler is diagrammed in Fig. 8-3a. Three processes are now needed: one for the input, a second for the buffer, and a third to compute the responses. Both the input and buffer processes must run at high priority, so that the input can pass its data to the buffer as quickly as possible and thus be available to input external events as much of the time as possible. The code for a simple buffered interrupt handler is listed in Fig. 8-3b. This handler uses a simple intermediate

```
CHAN OF INT work,buffer:
PRI PAR
  PAR
    INT data:
    WHILE TRUE                                    --input
      SEQ
        in? data
        buffer! data

    INT data:
    WHILE TRUE                                    --buffer
      SEQ
        buffer? data
        work! data

  INT data:
  WHILE TRUE                                      --work
    SEQ
      work? data
      handler(data)
```

Figure 8-3b
Code for a simple buffered interrupt handler

Chapter 8   Real-Time Processing

```
[buffer.size]CHAN OF INT buffer:
PRI PAR
  PAR
    INT data:                                      --input
    WHILE TRUE
      SEQ
        in? data
        buffer[0]! data

    PAR j=0 FOR buffer.size-1                      --buffer chain
      INT data:
      WHILE TRUE
        SEQ
          buffer[j]?    data
          buffer[j+1]!  data

  INT data:                                        --work
  WHILE TRUE
    SEQ
      buffer[buffer.size-1]? data
      handler(data)
```

Figure 8-3c
Code for a parallel buffered interrupt handler

buffer which can store only one value. This approach can easily be extended to larger buffers. Figure 8-3c lists another interrupt handler with a replicated PAR, each of whose processes can buffer another value. By adjusting the value of buffer.size, we can make the buffer as long as we wish.

This routine buffers data effectively because each individual buffer process will move its data whenever possible, ensuring that the work process will always be able to get work when it needs it. At the same time, the first process in the chain can accept data from the input whenever data arrives. This structure is really a very simple first-in, first-out buffer. However, the approach does have two significant drawbacks: first, every time a data element moves from one process to another, it is physically moved in the processor's memory; and second, significant context switching overhead is involved in the execution of the many independent processes needed for the buffer.

A FIFO Handler

A more efficient solution than the linear array of parallel processes is a first in, first out (FIFO) buffer queue, which can be made as large as necessary but does not require that data elements be copied from memory location to memory location as they move through the FIFO. Such a buffer will still provide storage for incoming signals from the input handler while passing the earliest signal to the computing process; however, it is much more complicated to program than the linear array of parallel processes.

The FIFO buffer must be able to satisfy two demands at once. First, it must be prepared to accept data from the input process at any time, and second, if there is data in the buffer, the buffer process must be able to pass the data to the compute process whenever the compute process is ready.

An elegant way to meet these requirements is described briefly in a paper by C. A. R. Hoare.† The program described creates a circular buffer which is input-driven both at the input of the buffer and at the output. Thus the work process must request data from the buffer rather than simply reading it in. A FIFO buffer program which implements this approach is listed in Fig. 8-4. In this program, the input routine is identical to the one listed earlier in Fig. 8-3b, while the output routine is almost the same. Rather than simply reading in data from the buffer, however, the output process in the FIFO program requests data via an additional channel called request. The processes and their channels are diagrammed in Fig. 8-5.

The FIFO routine begins with a declaration of three pointers, head, tail, and stop. These three pointers keep track of the data's position within the buffer: head points to the next element of the buffer to receive an input, tail points to the next element of the buffer to do an output, and stop points to the element in the buffer directly behind tail. A buffer using these pointers is illustrated in Fig. 8-6. The buffer itself can be thought of as a circular buffer in which tail is always chasing head. When data is read in from the input process, head advances; when data is passed to the computing process, tail advances. If head advances all of the way to stop, the buffer is full and no more data can be read in until a value is output and tail advances by one element. If tail catches up with head, the buffer is empty and no data can be output until a value is read in and head advances by one element. In this diagram, data element five is the most recent input, while data element zero is the next to be output.

Following the pointer declarations are the buffer array itself, fifo, defined to be of size buffer.size, and a useful constant, max, which is one less than buffer.size. Once these variables are initialized, the routine enters an infinite loop containing a PRI ALT.

This PRI ALT has two input sections, both guarded. The first ALT clause has priority and is for the input, which can proceed as long as there is room in fifo. This will be the case as long as the head pointer is not equal to stop. If an input does take place, the head pointer is incremented. Because fifo is a circular buffer, the increment operation is done modulo the buffer size, and for efficiency, performed by a comparison with max. The second ALT clause will only receive the request for an output if head is not equal to tail, indicating that fifo is empty. Any input is received on channel request. Once a request is

---

†. C. A. R. Hoare, "Communicating Sequential Processes," *Communications of the ACM*, vol. 21, 8 (August, 1978), p. 673.

```
CHAN OF INT work,buffer,request:
PRI PAR
  PAR
    INT data:
    WHILE TRUE                                          --input
      SEQ
        in?     data
        buffer! data

--         ------------   FIFO buffer begins here   -------------
    INT head,tail,stop:
    [buffer.size]INT fifo:
    VAL INT max IS buffer.size-1:
    SEQ
      head:=0                                           --initialize
      tail:=0
      stop:=buffer.size-1
      WHILE TRUE                                        --loop forever
        PRI ALT
          (head<>stop) & buffer? fifo[head]             --get input
            IF                                          --increment head
              head=max
                head:=0
              TRUE
                head:=head+1
          INT junk:
          (head<>tail) & request? junk                  --send work
            SEQ
              work! fifo[tail]                          --output data
              stop:=tail                                --increment stop
              IF                                        --increment tail
                tail=max
                  tail:=0
                TRUE
                  tail:=tail+1
--         ------------   FIFO buffer ends here    -------------
  INT data:                                             --work
  SEQ
    WHILE TRUE
      SEQ
        request! 0
        work? data
        handler(data)
```

Figure 8-4
Code for an interrupt handler with a request-driven circular FIFO buffer

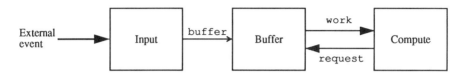

Figure 8-5
A FIFO buffered interrupt handler with data requests

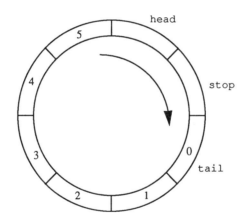

Figure 8-6
A circular buffer with the head pointer chasing the tail pointer

received, the data element that tail points to is output, and both stop and tail are incremented.

It may be inconvenient for some applications to request data from a buffer. A program that more closely models the parallel buffer discussed earlier does not require the output to request data, but instead simply reads it in just as is shown in Fig. 8-3a. If this is to happen, the input of new data must run in parallel with the output of old data, since ALT structures cannot use outputs. A circular FIFO buffer with this structure is diagrammed in Fig. 8-7 and listed in Fig. 8-8. The work and input processes for this FIFO buffer are identical to those used for the parallel buffer and are not repeated.

Three parts make up the FIFO buffer: a setup section followed by two parallel processes, one handling input from the input process and the other passing data to the computing process. Thick gray lines illustrate the control flow in Fig. 8-7, while the channels which communicate with each process are drawn with thin, dark lines.

The second FIFO buffer routine begins with the same declarations as those found in the first routine, with the addition of a boolean variable continue, and an interprocessor channel ready. First, the channel ready is declared, followed by a group of variables, and then the buffer (fifo) itself. The buffer size is again buffer.size, while max is one smaller. As before, the variable head always points to the element of the FIFO array in which any input data will be stored, while tail always points to the element of the FIFO array which must be output to the computing process. If head and tail point to the same element, the FIFO is

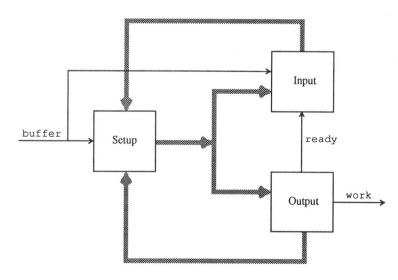

Figure 8-7
FIFO buffer process control flow and structure

```
CHAN OF BOOL ready:
BOOL continue:
INT head,tail,output.value,stop:
[buffer.size]INT fifo:
VAL INT max IS buffer.size-1:
SEQ
  head:=0
  tail:=0
  stop:=buffer.size-1
  WHILE TRUE
    SEQ
      continue:=TRUE                                        --set up
      IF
        head=tail                                      --empty buffer
          buffer? output.value
        TRUE                                         --buffer with data
          SEQ
            output.value:=fifo[tail]                  --assign output
            stop:=tail                                    --increment
            IF
              tail=max
                tail:=0
              TRUE
                tail:=tail+1
      PAR                              --do output and input in PAR
        WHILE continue                                        --input
          PRI ALT
            (head<>stop) & buffer? fifo[head]
              SEQ
                IF                                        --increment
                  head=max
                    head:=0
                  TRUE
                    head:=head+1
            ready? continue                              --quit PAR
              SKIP
        SEQ                                                 --output
          work! output.value
          ready! FALSE                                   --quit PAR
```

Figure 8-8
Code for a FIFO buffer process without an output request

empty. Initially, of course, the buffer is empty, and so head and tail are initialized to the same value, zero. The value stop points to the element just before tail, the last element in which new data can be stored before data already in the buffer will be overwritten.

After the variable initialization is complete, the buffer process enters an infinite loop. In this loop, the variable continue is first set to TRUE. The process then determines whether fifo is empty by comparing the values of head and tail. If head and tail are equal, the buffer is empty and the routine simply inputs the output.value from the buffer channel. If the buffer is not empty, output.value is assigned to the data element that tail points to, which is the next data element to be passed to the compute process. Stop is then updated, followed by tail. Since fifo is a circular buffer, tail cannot simply be incremented, but must be tested first to see if it is equal to max, and, if it is equal, set to point to the buffer element zero.

After output.value is assigned, the FIFO routine enters two parallel processes. The output process simply attempts to send output.value on the work channel, and, when it is successful, passes a signal to the input process on channel ready. The input process takes care of any data passed from channel buffer. This process waits in a loop controlled by continue, repeatedly waiting for an input from either buffer or ready. The buffer input is guarded so that no input can proceed if the FIFO is full (head <> stop is FALSE). Whenever data arrives on buffer, the value is stored in the fifo element pointed to by head, and then head is updated. As long as head is not equal to stop, the input process awaits another input on either buffer or ready. If the FIFO is full, the input process waits for a signal from the output process on channel ready indicating that the output process is also finished. When the ready signal arrives, continue becomes FALSE and both the input and output processes are finished.

When both the input and output processes are finished, the FIFO routine returns to setup mode and assigns or waits for a new output.value. The program's control flow is thus regulated by the output of data to the compute process.

Performance Comparisons

The performance of the interrupt handlers discussed above has been tested and the results are graphed in Figs. 8-9 and 8-10. For this series of tests, 1000 external events attempted to signal each interrupt handler at uniformly distributed random intervals of less than 20 milliseconds, and at an average rate of 10 milliseconds per event. For each test, the total time each handler delayed the input of interrupts was measured. The interrupting process executed on one processor and attempted to interrupt a second processor on which the input, buffer, and work routines were running. The processing time required for each interrupt was varied from 0 to 10 milliseconds. This processing time cannot be larger than the average event period, or the handler will never be able to keep pace with the external events. Note that this test examines only one criterion by which the performance of a handler might be measured. Other criteria could be used. For example, we might measure the average time between the input of an interrupt and the processing of that interrupt, that is, the time the event stays in the buffer. Our objective here, however, is to compare programming techniques rather than to examine queueing theory.

As we might expect, Fig. 8-9 shows that the simplest program performs very poorly, even when the time needed for processing is small compared to the average period of the external events. Simply placing the input in parallel with the computing improves the performance substantially. In the parallel case, the handler is available nearly all of the time, as long as the processing time is less than about half the period of the external events. The handler using a single buffer storing one event does better yet, providing good performance for computing times less than about 70% of the event period.

When FIFO buffers are used, the system performance improves dramatically. It is interesting to note in reference to Fig. 8-10 that the performance of both the FIFO routine using a linear chain of processes and the FIFO routine using a circular buffer without data requests is identical. (The scale of Fig. 8-10 is different from that of Fig. 8-9.)

The smallest possible circular FIFO buffer size is two, and the results show that, even though both of the FIFO routines are more complex than the single buffer routine, a FIFO of length two provides significantly better performance than the single buffer of length one. As the buffer length increases, the performance of both handlers continues to increase. When the buffer length is five and the processing time less than 80% of the event period, all of the events are processed with no delay. A buffer length of 10 can handle processing times up to 90% of the event period, while a 15-element buffer is not likely to delay any events for processing times up to 95% of the external event period.

For buffers with a length of 15 elements or less, the performance of the linear and circular (with no output request) FIFO routines is virtually identical. But as the buffer size increases, additional tests of the two routines with larger buffer lengths show that the circular buffer approach is more efficient. If the two routines are tested with a buffer length of 25, the circular buffer approach is nearly three times as fast as the buffer composed of a linear array of parallel processes. This effect is easily explained when we remember that the circular buffer does not move the data elements in memory as the data moves through the buffer. Also, the linear buffer has the overhead of many parallel processes to support. Only when the size of the buffer is larger than 15 do these effects become significant. So for small buffers, the much simpler program using a linear array of parallel processes is very suitable. (We should remember, however, that the buffer routines are only storing four-byte values. If larger amounts of data are stored, the overhead of physically moving data in memory will become more significant.)

It is also interesting to compare the performance of the two circular FIFO buffers. Although the buffer technique using data requests from the work process is shorter and more elegant, tests with a variety of buffer sizes and amounts of work

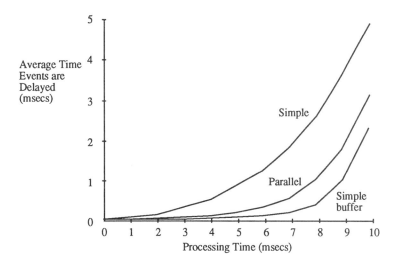

Figure 8-9
Total time for which various interrupt handlers delay input of events

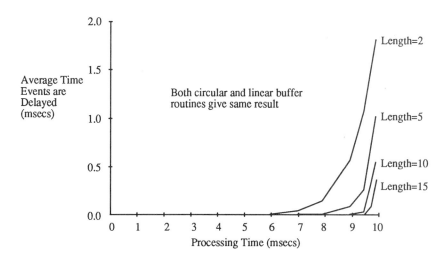

Figure 8-10
Total time the buffered interrupt handler with FIFOs of various sizes delays input of events

show that, when there is a significant amount of interrupt delay, the routine using data requests is 10% to 15% slower.

There is yet another point regarding interrupt handler performance which may be significant in some applications. It is obvious that the FIFO buffers could just as easily input directly from the interrupting process as from the input process. The input process really only provides an initial buffer, which could be included in the FIFO if we make the FIFO buffer one element larger. This input process is a very small, efficient loop, but using it does result in the overhead of an additional process. A test comparing each of the FIFO buffer routines shows that using the small, fast buffer at the front end does in fact improve the performance, but only very slightly. Programs including the input process typically execute slightly less than 0.5% faster than those which do not include the input process.

In Summary

Real-time processing can be done very effectively with transputer-based parallel systems. The effective use of real-time systems requires an efficient interrupt handler capable of receiving events from the real world with a minimum of delay. A variety of interrupt handlers ranging from the very simple to a parallel, buffered handler with three processes can be used.

Test results comparing the performance of these interrupt handlers show clearly that a simple, three-process interrupt handler can provide very good performance if it includes a simple FIFO buffer, even quite a small one. If only small buffer lengths are required, a simple linear array of buffer processes is adequate. If buffers larger than 15 are needed, circular buffers without data request channels should be used. For applications which require a relatively short processing time, a simple parallel handler, with or without a buffer, gives quite respectable performance. For applications which require a relatively long processing time, handlers with a FIFO buffer clearly provide much better performance.

Bibliography

The best sources of information about occam and transputer-based computer systems are the proceedings of the various user groups concerned with the application and investigation of occam and transputers. Both the European occam User's Group (OUG) and the North American Transputer Users Group (NATUG) have published proceedings. The conference proceedings listed below are in the order in which the conferences were held.

Muntean, Traian, ed., *Parallel Programming of Transputer Based Machines*, OUG-7, Proceedings of the 7th occam User Group Technical Meeting, 14-16 September, 1987, Grenoble, France (Amsterdam, The Netherlands: IOS, 1988).

Kerridge, Jon, ed., *Proceedings of the 8th Technical Meeting of the Occam User Group*, Sheffield City Polytechnic, U.K., March, 1988.

Askew, Charlie, ed., *Occam and the Transputer-Research and Applications*, OUG-9, Proceedings of the 9th occam User Group Technical Meeting, 19-21 September, 1988, Southampton, U.K. (Amsterdam, The Netherlands: IOS, 1988).

Bakkers, Andre, ed., *Applying Transputer Based Parallel Machines*, OUG-10, Proceedings of the 10th occam User Group Technical Meeting, 3-5 April, 1989, Enschede, Netherlands (Amsterdam, The Netherlands: IOS, 1989).

Wexler, J., ed., *Developing Transputer Applications*, OUG-11, Proceedings of the 11th occam User Group Technical Meeting, 25-26 September, 1989, Edinburgh, U.K. (Amsterdam, The Netherlands: IOS, 1989).

Turner, Stephen J., ed., *Tools and Techniques for Transputer Applications*, OUG-12, Proceedings of the 12th occam User Group Technical Meeting, 2-4 April, 1990, Exeter, U.K. (Amsterdam, The Netherlands: IOS, 1990). Includes an in-depth processor farm analysis, p. 179, R.W.S. Tregidgo and A.C. Downton.

Stiles, G. S., ed., *NATUG 1*, Proceedings of the First Conference of the North American Transputer Users Group, Salt Lake City, Utah, April 5-6, 1989.

Board, John A., ed., *NATUG 2*, Proceedings of the Second Conference of the North American Transputer Users Group, Durham, North Carolina, Oct 18-19, 1989. Includes an in-depth discussion of deadlock-free packet networks, p. 139, Martin Shumway.

Bibliography

Wagner, Alan S., ed., *Transputer Research and Applications 3*, NATUG-3, Proceedings of the Third Conference of the North American Transputer Users Group, April 26-27, 1990, Sunnyvale, California, (Amsterdam, The Netherlands: IOS, 1990).

Freeman, Len and Chris Phillips, eds., *Applications of Transputers 1*, Proceedings of the first Applications of Transputer conference, August 23-25, 1989, Liverpool, U.K. (Amsterdam, The Netherlands: IOS, 1989).

A second excellent source of information is the documentation and application notes published by INMOS.

INMOS Limited, *Communicating Process Architecture*, Prentice Hall International Series in Computer Science, Hemel Hempstead, U.K.: Prentice Hall International, 1988.

INMOS Limited, *occam Programming Manual*, Prentice Hall International Series in Computer Science, Englewood Cliffs NJ: Prentice Hall International, 1984.

INMOS Limited, *occam 2 Reference Manual*, Prentice Hall International Series in Computer Science, Hemel Hempstead, U.K.: Prentice Hall International (U.K.) Ltd., 1988.

INMOS Limited, *The Transputer DataBook*, Berkeley, CA: Consolidated Printers, 2nd ed., 1989.

INMOS Limited, *Transputer Development System*, Hemel Hempstead, U.K.: Prentice Hall International (U.K.) Ltd., 1988.

INMOS Limited, *Transputer Instruction Set*, Hemel Hempstead, U. K.: Prentice Hall International (U.K.) Ltd., 1988.

INMOS Limited, *Transputer Reference Manual*, Hemel Hempstead, U.K.: Prentice Hall International (U.K.) Ltd., 1988.

INMOS Limited, *Transputer Technical Notes*, Hemel Hempstead, U.K.: Prentice Hall International (U.K.) Ltd., 1988.

A third source of information is the general literature dealing with parallel processing, especially distributed-memory computing systems, transputers, and occam.

Bibliography

Axford, T., *Concurrent Programming: Fundamental Techniques for Real-Time and Parallel Software Design*, New York NY: John Wiley and Sons, Inc., 1989.

Ben-Ari, M., *Principles of Concurrent and Distributed Programming*, Englewood Cliffs, NJ: Prentice Hall, 1989.

Bertsekas, D. P., and J. N. Tsitsiklis, *Parallel and Distributed Computation*, Englewood Cliffs, NJ: Prentice Hall, 1989.

Burns, A., *Programming in occam 2*, London, U.K.: Addison Wesley Publishing Company, 1988.

Carling, A., *Parallel Processing - The Transputer and occam*, New York, NY: John Wiley and Sons, Inc., 1988.

Crichlow, J., *An Introduction to Distributed and Parallel Computing*, Englewood Cliffs, NJ: Prentice Hall, 1988.

Desrochers, George R., *Principles of Parallel and Multiprocessing*, New York, NY: McGraw Hill Book Co., 1987.

Hoare, C. A. R., *Communicating Sequential Processes*, Hemel Hempstead, U.K.: Prentice Hall International, 1985.

Hockney, R. W. and C. R. Jesshope, *Parallel Computers 2: Architecture, Programming and Algorithms*, Bristol U.K.: IOP Publishing Ltd., 1988.

Hwang, Kai, and Faye A. Briggs, *Computer Architecture and Parallel Processing*, New York, NY: McGraw Hill Book Co., 1984.

Jones, G. and M. Goldsmith, *Programming in occam 2*, Hemel Hempstead, U.K.: Prentice Hall International, 1988.

Pountain, Dick and David May, *A Tutorial Introduction to occam Programming*, Blackwell Scientific Publications Professional Books, London U.K., 1987.

Wexler, John, *Concurrent Programming in occam 2*, Ellis Horwood Series in Computers & Their Applications. Chichester, U.K.: Ellis Horwood Limited, 1989.

Index

! 14
? 14

A

Abbreviations 13
Activity diagram
   double buffering 100
   loading
      bidirectional all the way 124
      bidirectional halfway 124
      fast pipeline 121
      simple 121
      small buffers 123
   pipeline 91
   single buffering 98
   triple buffering 101
AFTER 16
Algorithm
   routing 147, 149
      dynamic 147
Algorithmic limitations 31
Algorithmic parallelism 89
ALT 14, 15
   guard 15
   PRI 15
   replication 16
AND 17
Arbitrary network 175, 178, 193
   program 175
Architecture 1
   classes 23
   data parallelism 113

   equivalence 32
   history 23
   irregular 35
   limitations 31
   processor farm 57
   regular 35
   scaling 35
   special-purpose 32
Array 11
   channels 14
Assignment 11, 12

B

Background task 213
Bandwidth 24, 28
   shared memory 26
Bidirectional loading 124
BOOL 17
Booting a transputer 183
   arguments 184
   extra-copy method 183
   from link 39, 183, 184, 190
   from memory 39
   parallel worm 195
   self-copy method 183
Bootstrap program 183
Breadth-first search 176, 181, 185, 199
   toroid 181
Buffer
   circular 218, 223, 224
      performance comparison 224
   double 99

Index

activity diagram 100
code 99
efficiency 105
loading 121
    performance 127
matrix multiplication 109
performance 105
efficiency 105
first-in, first-out 217
interrupt handler 216
multidimensional 111
performance 103
single 98
    activity diagram 98
    code 98
    performance 105
size for loading 123
test 103
triple 101
    activity diagram 101
    code 101
    efficiency 106
    performance 105
when to use 105

Bus 26

C 20

C

CASE 18

CHAN 14

CHAN OF ANY 14

Channel
ALT 15
array 14
communication 14
configuration 36
definition 14
independence 156
input 14, 15
multiplexer 150, 154
    bidirectional 160, 163

OF ANY 14
output 14
point-to-point 28
processor farm 57
TIMER 14
tree 76
type 14
undefined type 14
virtual 149, 156, 158
    one-way 150
    two-way 160

Child
processor farm 64
worm 176
    exploratory 204
    parallel 193, 197
    sequential 187, 190, 192

Circular buffer 218, 223, 224
performance comparison 224

Comment
folds 11
program 11

Communicating data sets 137
efficiency 142
memory requirements 144

Communication
bandwidth 24
bottleneck 23, 24
channel size 24
concurrency 6
cost of 31
efficiency 130
    simple load 121
expanding data sets 130
handler 152, 154
    code 153
    diagram 152
in parallel with computation 105
interprocessor 146
limiting performance 106
loading 119
multiple channels 24, 25

232   *Parallel Programs for the Transputer*

overhead 4, 6, 31, 91
  pipeline 94, 96
parallel processes 13
pipeline 90, 91, 96
point-to-point 28
priority 13, 66
processor farm 57, 70
routing 147
shared memory 26
shell 146
single-instruction, single-data 26
speed of light 24

Configuration
  channel 36
  description 36
  hypercube 47
  irregular network 39
  links 37
  PLACE 37
  PLACED PAR 37
  processor farm 72
  real-time 30
  ring 42
  single-processor 37
  toroid 44
  tree 51
  two-processor 38

Context switching 8

Control structures 16

Controller, farm 59

Convolution 4, 130, 131, 137, 142
  efficiency 6
    example 4
  performance 6

D

Data distribution 91, 115
  code 116
  example 115
  pipeline 90
  ring 114, 118

Data flow 89

Data parallelism 89, 113, 145
  architecture 113
  communicating data sets 137
  convolution 130
  data distribution 115
  efficiency 115, 142
  expanding data sets 130, 133
    performance 135
  flexibility 114
  frequency transforms 114
  loading 119
  overhead 114
  program distribution 113
  sampling 128
  scalability 114
  shifting 137
  trade-off 114

David xii

Deadlock 19, 147

Deadlock-free network 150

Debugger 20, 175

Debugging 19, 20
  deadlock 19
  livelock 19

Depth-first search 176
  toroid 181

Distributed memory 28
  architectural equivalence 32
  interconnections 30
  switching 30

Distribution of program 89, 91

E

Editor 11

Efficiency 2–8, 106
  architecture 32
  buffer 105
  communication 4, 130
  comparison, shift vs. expand 142

Index

convolution 4
data distribution 115
definition 2
double buffer 105
expanding data sets 135
granularity 63
hardware 8
load balancing 61
loading 128
    bidirectional 124
    fast pipeline 121, 123
maximum pipeline 94
overhead 6
parallel computer 4
pipeline 91, 93, 94, 96, 105, 115
pipeline buffering 103
processor farm 69, 83, 85
relative 2, 4
shifting 141
simple processor farm 60
speedup 3
superlinear 3
task distribution 60, 61
task variability 61
theoretical limit 93

ELSE 18

Expanding data sets 130
    efficiency 135, 142
    memory requirements 136, 144
    one-dimensional 131
        code 131
    performance 135
    two-dimensional 131
        code 133

Exploratory worm 199
    child 204
    code 200
    code size 211
    input & output routines 204
    kill 206
    link failure 211
    parent 200, 204

    performance 211
    root node 210
    test state 206

F

FALSE 17
Farm, processor 57
FIFO see first-in, first-out buffer 217
Fine grain 63
First-in, first-out buffer 163, 217
    diagram 221
    performance 224
Folds 11
    comments 11
Font for code listings 11
FORTRAN 20
Frequency transforms 114
FUNCTION 19
Functions 19

G

Geometric parallelism 113
Grammarian xii
Granularity 63, 83, 85, 86, 113
    efficiency 63
    fine grain 63
    large grain 63
    medium grain 63
    transputer 63
Guard 15, 17, 18, 223

H

Hypercube 47, 113
    configuration 47
    diagram 47
    interprocessor distance 51
    network 47
    scaling 50

toroid as subset 51

I

IF 17
   guard 17
Indentation 11, 12
Input 14, 15
   AFTER 16
   guard 18
INT 11
Interprocessor communication 146
Interprocessor distance
   hypercube 51
   ring 43
   toroid 46
   tree 55
Interrupt
   fixed interval 213
   random interval 213
Interrupt handler 214
   buffer 216
      code 216
      diagram 216
   delay 213
   first-in, first-out buffer
      diagram 221
   parallel 215
      code 215
      diagram 215
   performance comparisons 223
   processing rate 213
   simple 214
      code 214
      diagram 214
Irregular architecture 35
Irregular network 35, 39
   configuration 39
   diagram 39
   message-passing 145
   real-time processing 213

L

Large grain 63
Large processor farm 63
Latency 101
   pipeline 93, 99
Library 19
Limit of efficiency
   pipeline 93
Linear array 89
Link 13, 36
   address assignment 37
   bandwidth 8
   bidirectional pipe 106
   booting 39, 184, 190
   diagram 37
   efficiency
      fast pipeline load 121
      simple load 121
   failure 211
   hardware 8, 36
   interconnections 13
   multiplex 152
   network construction 36
   peek and poke 200
   real-time processing 213
   switching 36
   testing 188, 197, 200
   transputer 8
   virtual channels 149
Livelock 19
Load balancing 61, 113
Loading 119
   bidirectional 124
      code 124
      efficiency 124
      performance 128
   bidirectional all the way
      activity diagram 124
   bidirectional halfway
      activity diagram 124

advantage 127
buffer size 123
double buffer performance 127
fast pipeline 121
   activity diagram 121
   code 121
   efficiency 121, 123
   performance 128
memory organization 127
performance 127
programs 175
simple 119
   activity diagram 121
   code 119
   efficiency 121
   performance 127
small buffers
   activity diagram 123

Logical variables 17

M

Mandelbrot set 70
   calculation 72
      efficiency 83
   function 70

Matrix multiplication 108
   code 108
   double buffering 109

Medium grain 63
   transputer 63

Memory requirements
   data parallelism 144

Memory testing 188

Message passing 28, 145
   delivery 146
   portability 146
   processor farm 145
   structure 146
   traffic 147
   unpredictability 146

Multidimensional pipeline 106

Multiple-instruction, multiple-data 26, 28, 35, 57, 113
   architectural equivalence 32
   complexity 28
   hardware 29
   scalability 29

Multiple-instruction, single-data 26

Multiplex channel 154, 158
   bidirectional 160, 163

Multiplex channels 150

Multiplex link 152

N

Network 39
   arbitrary 175, 178, 193
   construction 36
   deadlock 148
   deadlock-free 150
   emphasis in book 1
   failures 199
   grid 108
   hypercube 47
   interconnections 13
   irregular 35, 39
      message-passing 145
      real-time processing 213
      used for 36
   limitations 31
   loading programs 175
   number of links 36
   overhead 31
   processor farm 57
   programming examples 8
   regular 35, 175, 176
      advantages 35
      used for 36
   ring 8, 42, 89
   root node 39
   scaling 35
   searching strategy 176

shared memory 12
single-processor 37
switched 1, 30, 64
test 175
testing 199
toroid 8, 44, 178, 181
tree 51, 176
two-processor 38
unknown 199
unswitched 1

Node controller
  processor farm
    code 81
    diagram 80

NOT 17

O

occam 11–19
  ! 14
  ? 14
  abbreviations 13
  AFTER 16
  ALT 14
  AND 17
  array 11
  assignment 12
  BOOL 17
  CASE 18
  CHAN 14
  channel 14
  comments 11
  control 16
  ELSE 18
  FALSE 17
  folds 11
  font 11
  FUNCTION 19
  functions 19
  guard 18
  IF 17
  indentation 11, 12
  input 14
  guard 18
  library 19
  logical variables 17
  NOT 17
  OR 17
  output 14
  PAR 12
  PLACED PAR 13
  PRI ALT 15
  PRI PAR 13, 15
  PROC 18
  procedures 18
  process 11
  real-time 13
  real-time processing 213
  replication 16
  SEQ 12
  shared memory 12
  TIMER 14, 15
  TRUE 17
  variables 11
  WHILE 16

Occam, William of 2

Operating systems, parallel 21, 146

OR 17

Output 14

P

PAR 12
  replication 16

Parallel processing
  limits 63

Parallel processors
  algorithmic limitations 31
  complexity 28
  distributed memory 28
  limitations 31
  multiple-instruction, multiple-data 26, 28
  multiple-instruction, single-data 26
  scalability 29

Index

shared memory 26
single-instruction, multiple-data 25
single-instruction, single-data 25
Parallel programs
  overhead 4
Parallel search 181
Parallel worm 193
  booting 195
  child 193, 197
  code 193
  parent 193, 195
  performance 199
  root node 197
  tree 197
Parent
  processor farm 64
  worm 176
    exploratory 200, 204
    parallel 193, 195
    sequential 187, 192
Pascal 20
Performance
  buffer 105
  circular buffers 224
  communication limited 106
  expanding data sets 135
  first-in, first-out buffer 224
  interrupt handler 223
  limit 4
  loading 127
    double buffer 127
    simple 127
  measurements 10, 14
  pipeline 94, 103
    theoretical limit 96
  pipeline buffering 103
  processor farm 83
  router 172
    four-way 173
    one-way 172
    two-way

    performance 172
  shifting 141
  worm
    exploratory 211
    parallel 199
    sequential 193
Pipe 89
Pipeline
  activity diagram 91
  advantages 89
  buffer
    double 99, 105
    single 98, 105
    test 103
    triple 101, 105
    when to use 105
  communication 90, 91, 96
    limited 106
    overhead 94, 96
  data distribution 90, 91
  definition 89
  disadvantages 89
  efficiency 91, 93, 94, 96, 103, 115
  empty 93
  example 94
  fill 93
  flexibility 113
  latency 93, 99, 101
  limit of efficiency 93, 94
  loading 121
  matrix multiplication 108
  maximum efficiency 94
  multidimensional 106
  overhead 91
  packet size 91, 93, 94, 96
  parallelism 145
  performance 94, 96, 103, 105
  processing
    real-time 106
    total time 93
  program 89
    distribution 113

flexibility 89
program distribution 91
ring 89
scalability 90
two-dimensional 109

PLACE 37

PLACED PAR 13, 37

Portable programs 35

PRI ALT 15

PRI PAR 13, 15

PROC 18

Procedures 18

Process 11

Processor farm 57, 193
architecture 57
child 64, 76
communication overhead 70
configuration 72
control 57
controller 59, 80
  code 81
definition 57
diagram 64
efficiency 69, 83, 85
example 70
granularity 83, 85, 86
interprocessor distance 64
large 63
Mandelbrot set 70
message passing 145
parallel processes 66
parent 64
performance 83
root node 59
simple 59
  efficiency 60
size calculation 69
storage 69
structure 57
task queue 70

tree 72, 76
workers 59
worm 175

Program 1
arbitrary networks 175
comments 11
distribution 89, 91, 93, 114
examples 1
folds 11
font listed in 11
indentation 11
pipeline 89
portability 146
portable 35
scalable 35

Program editor 11

Q

Queue
processor farm 70
real-time 217

R

Random events 213

Real-time processing 13, 213
definition 213
occam 213
pipeline 106
timing 14

Regular architecture 35
advantages 35

Replication 16
worm 176

Ring 42, 113
bidirectional loading 124
  performance 128
configuration 42
data distribution 114, 118
diagram 42
expanding data sets 131

Index

interprocessor distance 43
loading performance 127
network 8, 42
pipeline 89
program tasks 114
router 149
  one-way 156
    performance 172
  two-way 163
    performance 172
scaling 43
subset of toroid 47
Root node 39
  exploratory worm 199, 210
  parallel worm 197
  processor farm 59, 80
  sequential worm 193
Router 145
  algorithm 147, 149
  bidirectional 160
  channel independence 156
  deadlock 147, 148
  deadlock-free 150
  dynamic 147
  first-in, first-out 163
  four-way 165, 168
    performance 173
  message distribution 172
  one-way 150, 156
    performance 172
  performance 172
  ring 149, 163
  simple
    code 157
  toroid 165
  two-way 163
    performance 172

S

Sampling 128
  code 128
Scalability

multiple-instruction, multiple-data 29
parallel processors 29
shared memory 26
single-instruction, multiple-data 29
switched networks 30
Scaling
  architectures 35
  data parallelism 114
  hypercube 50
  pipeline 90
  programs 35
  ring network 43
  toroid 47
  tree 55
Scope of variables 11
Searching strategy 176
  breadth-first 176, 185, 199
    search 181
  depth-first 176
  exploring 199
  parallel 181
  sequential 181
SEQ 12
  replication 16
Sequential search 181
Sequential worm 185
  child 187, 190, 192
  code 185
  parent 187, 192
  performance 193
  root node 193
Shared memory 12, 26
  architectural equivalence 32
  bandwidth 26, 28
  bus 26
  common memory 28
  communication channel 26
  control flow 27
  flexibility 29
  message passing 28

processor interaction 28
programs 27
scalability 26
single-instruction, single-data 28

Shifting 137
code 137
efficiency 141, 142
memory requirements 144
performance 141

Simple processor farm 59
efficiency 60

Single-instruction, multiple-data 25, 63, 113
communication 26
control flow 27
cpu 25
scalability 29

Single-instruction, single-data 25
architectural equivalence 32
shared memory 28
supercomputers 26

Single processor
configuration 37
network 37

Speedup 3
superlinear 3

Storage
processor farm 69

Sueanne xii

Supercomputer 26

Superlinear speedup 3
purported examples 3

Switched network 64

Switched networks 30
expansibility 30

Switching 64
links 36

T

T800 8

Task distribution 60, 61, 89

Task granularity 63

Task queue 70

Task variability 61

Tester 175

Testing
link 188, 197
memory 188
networks 199
processor 190

Text editor 11

TIMER 10, 14, 15

Timing performance 10

Toroid 44, 113, 130, 137, 178, 181
advantages 46
breadth-first 181
configuration 44
depth-first search 181
diagram 44
interprocessor distance 46
network 8, 44
ring as subset 47
router 165
performance 173
scaling 47
subset of hypercube 51

Transputer 1, 8–10
architectures 1
booting
link 39
memory 39
context switching 8
diagram 8
efficiency 8
hardware 8
link 8, 36
medium grain 63

Index

multiple-instruction, multiple-data 28
network construction 36
network for examples 8
processor farm 57
program 1
programming language 11
real-time processing 213
T800 8
variety of techniques 1

Tree 51, 185
  as a subset 55
  child 176, 187, 192
  configuration 51
  descendant 190
  diagram 51
  interprocessor distance 55
  network 51, 176
  parallel worm 193, 197
  parent 176, 187
  processor farm 72, 76
  scaling 55

triple buffer 106

TRUE 17

Two processor
  configuration 38
  diagram 38
  network 38

V

Variables
  array 11
  logical 17
  occam 11
  scope 11

Vector computer 26

Virtual channel 149, 156, 158
  one-way 150
  two-way 160

von Neumann bottleneck 23, 25

W

WHILE 16

William of Occam 2

Workers
  processor farm 59

Worm 175
  bootstrap 183, 185
  child 176
  definition 175
  descendant 190
  exploratory 199
  parallel 193
  parent 176
  performance 193, 199, 211
  processor farm 175
  replication 176
  searching strategy 176
  sequential 185

Parallel Programs for the Transputer

Ronald S. Cok

The transputer, a single-chip microcomputer, is an increasingly popular choice in the rapidly growing technology of parallel computing. Transputers offer a complete parallel computing environment, including both hardware and software solutions. **Parallel Programs for the Transputer** clearly illustrates the diverse techniques needed to construct and program a variety of parallel computers using the transputer.

Specific topics include:
- a variety of computer architectures
- six programming methodologies
- programming techniques ranging from simple to complex, complete with practical working examples and efficiency demonstrations.

After reading this book, you'll be thoroughly familiar with a broad range of transputer programming techniques, and the advantages, tradeoffs, and efficiency of each one.

PRENTICE HALL, Englewood Cliffs, N.J. 07632

ISBN 0-13-651480-4